Cognitive Systems Monographs

Volume 43

The Cognitive Systems Monographs (COSMOS) publish new developments and advances in the fields of cognitive systems research, rapidly and informally but with a high quality. The intent is to bridge cognitive brain science and biology with engineering disciplines. It covers all the technical contents, applications, and multidisciplinary aspects of cognitive systems, such as Bionics, System Analysis, System Modelling, System Design, Human Motion Understanding, Human Activity Understanding, Learning of Behaviour, Man-Machine Interaction, Smart and Cognitive Environments, Human and Computer Vision, Neuroinformatics, Humanoids, Biologically motivated systems and artefacts, Autonomous Systems, Linguistics, Sports Engineering, Computational Intelligence, Biosignal Processing, or Cognitive Materials—as well as the methodologies behind them. Within the scope of the series are monographs, lecture notes, selected contributions from specialized conferences and workshops, as well as selected Ph.D. theses.

Indexed by SCOPUS, DBLP, zbMATH, SCImago.

More information about this series at http://www.springer.com/series/8354

Lorenzo Magnani

Eco-Cognitive Computationalism

Cognitive Domestication of Ignorant Entities

Springer

Lorenzo Magnani
Department of Humanities,
Philosophy Section and Computational Philosophy
Laboratory
University of Pavia
Pavia, Italy

ISSN 1867-4925 ISSN 1867-4933 (electronic)
Cognitive Systems Monographs
ISBN 978-3-030-81446-5 ISBN 978-3-030-81447-2 (eBook)
https://doi.org/10.1007/978-3-030-81447-2

This Springer imprint is published by the registered company Springer Nature Switzerland AG
The registered company address is: Gewerbestrasse 11, 6330 Cham, Switzerland

If we are trying to produce an intelligent machine, and are following the human model as closely as we can, we should begin with a machine with very little capacity to carry out elaborate operations or to react in a disciplined manner to orders (taking the form of interference). Then by applying appropriate interference, mimicking education, we should hope to modify the machine until it could be relied on to produce definite reactions to certain commands. This would be the beginning of the process.

Alan Turing, Intelligent Machinery, 1948

To my cat Sheena

Preface

The book addresses a new approach to computation I call "eco-cognitive", which stresses the attention to the question of the overwhelming domestication of ignorant entities, which is at work in our current societies. I think the intellectual and didactic virtues of this approach resort to the conviction that seeing at computation in the special interdisciplinary perspective I am proposing avoids reducing the analysis of computation to a rigid one-sided view, privileging the theoretical commitment towards the study of computation as a variegated phenomenon that regards human beings in their individual and social lives.

The book aims at stressing that an open-minded analysis of computation must take into account:

- the *distributed* and *embodied* nature of computational tools, models, and methods, ultimately related to the idea of the current importance of the external cognitive machines as mediators of cognition;
- the central role of the dynamics of the production—and of the rational handling—of new kinds of unconventional computations, by referring to the various *multimodal* related epistemological aspects;
- the fact that computation can be satisfactorily examined only in an *eco-cognitive* perspective.

These topics are analyzed in terms of what I consider the main tenets of an eco-cognitive approach computation:

1. Chapter 1 The concepts of information, computation, and cognition are variously interpreted and explained and still lead to ambiguous results. I contend that seeing the evolutionary emergence in humans of information, meaning, and of the first kinds of cognition as the outcome of dynamic coevolutionary interactions between brain/mind internal processes, body itself, and external environment can be extremely useful (1) to clarify the most common misunderstandings and the basic vagueness of the concepts above and (2) to appropriately describe "computation" as an evolving concept subjected to continuous transformations of meaning. To this aim, I also take advantage of the dynamic concepts

of salience and pregnance derived from Thom's catastrophe theory. When the physical computation is seen in the perspective of the ecology of cognition, it is easy to understand Turing's original ideas concerning the emergence of information, cognition, and computation in organic, inorganic, and artefactual agents, which I also briefly illustrate in this chapter. I also show that seeing computation as dynamically active in distributed physical entities of various kinds suitably transformed so that data can be encoded and decoded to obtain appropriate results further sheds light on what I call *eco-cognitive computationalism*. I hope it will become clear that eco-cognitive computationalism does not aim at furnishing an ultimate and static definition of the concepts of information, cognition, and computation, such as a textbook could provide, instead, it intends, by respecting their historical and dynamical character, to propose an intellectual framework that depicts how we can understand their forms of "emergence" and the modification of their meanings.

2. Chapter 2. Eco-cognitive computationalism sees computation in context, exploiting the ideas developed in those projects that have originated the recent views on embodied, situated, and distributed cognition. As illustrated in the previous chapter, Turing's original intellectual perspective has already clearly depicted the evolutionary emergence in humans of information, meaning, and of the first rudimentary forms of cognition, as the result of a complex interplay and simultaneous coevolution, in time, of the states of brain/mind, body, and external environment. This cognitive process played a fundamental heuristic role in Turing's invention of the universal logical computing machine. It is by extending this eco-cognitive perspective that we can see that the recent emphasis on the simplification of cognitive and motor tasks generated in organic agents by *morphological aspects* implies the construction of appropriate *mimetic bodies*, able to render the accompanied computation simpler, according to a general appeal to the "simplexity" of animal embodied cognition. I hope it will become clear that, as already contended in the previous chapter, eco-cognitive computationalism does not aim at furnishing a final and fixed definition of the concept of computation, but stresses the historical and dynamical character of the concept.
3. Chapter 3. *Locked* and *unlocked strategies* are at the center of this chapter, as ways of shedding new light on the cognitive aspects of deep learning machines. The character and the role of these cognitive strategies, which are occurring both in humans and in computational machines, are indeed strictly related to the generation of cognitive outputs, which range from weak to strong levels of knowledge creativity. For a better understanding of the computational aspect, I will also illustrate the abductive character of the concept of *anticipation* in human cognition linking it to the cognitive problem of spatiality and of the genesis of space in the description of the abductive role of the *Abschattungen* (adumbrations), as they are described in the framework of the philosophical tradition of phenomenology. Anticipations share various features with visual and manipulative abduction and prove it to be a useful tool to favor the characterization of the two kinds of strategic reasoning I am introducing in this chapter. It will become clear that the differences between strong and weak

levels of creativity lead to important consequences when we analyze computational AI programs, such as AlphaGo/AlphaZero, which aim at performing various kinds of abductive hypothetical reasoning. In these cases, the programs are characterized by *locked* abductive strategies: they deal with weak (even if sometimes amazing) kinds of hypothetical creative reasoning, because they are limited in what I call *eco-cognitive openness*, which instead qualifies human cognizers who are performing higher kinds of abductive creative reasoning, where cognitive strategies are instead *unlocked*.

4. Chapter 4. In the first two chapters of this book, we have stressed that eco-cognitive computationalism sees computation in context, following some of the main tenets advanced by the recent cognitive science views on embodied, situated, and distributed cognition. We have also described the new attention in computer science devoted to the relevance in the computation of the morphological features. It is by further deepening and analyzing the perspective opened by these novel fascinating approach that we see ignorant bodies as *domesticated* to become useful "mimetic bodies" from a computational point of view. This new perspective shows how the computational domestication of ignorant entities can originate new variegated unconventional cognitive embodiments, so opening the new research field of the so-called *natural computing*. In the last part of the chapter, I will introduce the concept of overcomputationalism, showing that my proposed framework helps us to see the related concepts of pancognitivism, paninformationalism, and pancomputationalism in a more naturalized and prudent perspective, avoiding the excess of old-fashioned ontological or metaphysical overstatements. Finally, I will submit to the attention of the reader a question that in my opinion synthesizes all the problems described in the chapter: will the future of eco-cognitive settings be computationally tailored or humanly tailored?

I am grateful to many colleagues and friends for their helpful suggestions and much more. For valuable comments and discussions, I am indebted and grateful to John Woods, Paul Thagard, Ping Li, Atocha Aliseda, Woosuk Park, Nancy J. Nersessian, Giuseppe Longo, Yukio Oshawa, Akinori Abe, Michael Leyton, Dov Gabbay, John Josephson, Walter Carnielli, Gerhard Schurz, Balakrishnan Chandrasekaran, Jon Williamson, Douglas Walton, Cameron Shelley, Sami Paavola, Luís Moniz Pereira, Thomas Addis, Lorenza Magnani, Giovanna Magnani, Diderik Batens, Joke Meheus, Simon Colton, Athanassios Raftopoulos, Michael Hoffmann, Ilkka Niniluoto, Theo A. F. Kuipers, Chris Sinha, Ryan D. Tweney, Ferdinand D. Rivera, Peter Flach, Antony Kakas, Oliver Ray, Luis A. Pineda, Atsushi Shimojima, Pat Langley, Demetris P. Portides, and to my collaborators Selene Arfini and Alger Sans Pinillos. Also of great help was correspondence or conversation with several people whose influence on various of the book's particularities is more or less palpable. I especially wish to acknowledge Hanne Andersen, Otávio Bueno, Sanjay Chandrasekharan, Marcelo Dascal, Gordana Dodig-Crnkovic, Mauro Dorato, Michel Ghins, Marcello Guarini, Ricardo Gudwin, Albrecht Heeffer, Mireille Hildebrandt, Kenneth Einar Himma, Gerhard Minnameier, Margaret Morrison, Alfredo

Pereira, Ahti-Veikko Pietarinen, Dagmar Provijn, João Queiroz, Chiaki Sakama, Colin Schmidt, Nora Schwartz, Frederik Stjernfelt, Mauricio Suárez , Jeroen van den Hoven, Peter-Paul Verbeek, Riccardo Viale, and Marion Vorms. The preparation of the volume would not have been possible without the contribution of resources and facilities of the Computational Philosophy Laboratory (Department of Humanities, Philosophy Section, University of Pavia, Italy). Research for this book was supported by the PRIN 2017 Research 20173YP4N3—MIUR, Ministry of University and Research, Rome, Italy. This project was conceived as a whole, but as it developed, various parts have become articles, which have now been excerpted, revised, and integrated into the current text.

Parts of this book are excerpted from L. Magnani (2018), Computationalism in a dynamic and distributed eco-cognitive perspective, in G. Dodig-Crnkovic and M. Burgin (eds.), *Philosophy and Methodology of Information*, World Scientific, Singapore, pp. 365–288 (reprinted with permission) (Chapter one); L. Magnani (2018), Eco-cognitive computationalism. From mimetic minds to morphology-based enhancement of mimetic bodies, *Entropy*, 20(6), 430 http://dx.doi.org/10.3390/e20060430 (Chapter two); L. Magnani (2019), AlphaGo, locked strategies, eco-cognitive openness, in H. Zenil and S. Bringsjord, eds., "Philosophy and Epistemology of Deep Learning" a special issue of *Philosophies* 2019, 4(1), 8, https://doi.org/10.3390/philosophies4010008 (Chapter three); L. Magnani (2020), Computational domestication of ignorant entities. Unconventional cognitive embodiments, *Synthese*, Special Issue edited by L. Magnani and S. Arfini "Knowing the Unknown: Philosophical Perspectives on Ignorance", online first. (Chapter four). I am grateful to Springer and World Scientific for granting permission to include portions of previously published articles.

Pavia, Italy
January 2021

Lorenzo Magnani

Contents

Chapter 1
Computationalism in a Dynamic and Distributed Eco-Cognitive Perspective

The concepts of information, computation, and cognition are variously interpreted and explained and still lead to ambiguous results. I will contend that seeing the evolutionary emergence in humans of information, meaning, and of the first kinds of cognition as the outcome of dynamic coevolutionary interactions between brain/mind internal processes, body itself, and external environment can be extremely useful (1) to clarify the most common misunderstandings and the basic vagueness of the concepts above, and (2) to appropriately describe "computation" as an evolving concept subjected to continuous transformations of meaning. To this aim I will also take advantage of the dynamic concepts of salience and pregnance derived from Thom's catastrophe theory. When physical computation is seen in the perspective of the ecology of cognition it is easy to understand Turing's original ideas concerning the emergence of information, cognition, and computation in organic, inorganic, and artefactual agents, I will also briefly illustrate in this chapter. I will show that seeing computation as dynamically active in distributed physical entities of various kinds suitably transformed so that data can be encoded and decoded to obtain appropriate results further sheds light on what I call eco-cognitive computationalism. I hope it will become clear that eco-cognitive computationalism does not aim at furnishing an ultimate and static definition of the concepts of information, cognition, and computation, such as a textbook could provide, instead it intends, by respecting their historical and dynamical character, to propose an intellectual framework that depicts how we can understand their forms of "emergence" and the modification of their meanings.

L. Magnani, *Eco-Cognitive Computationalism*, Cognitive Systems Monographs 43,
https://doi.org/10.1007/978-3-030-81447-2_1

1.1 Introducing Eco-cognitive Computationalism

1.1.1 Eco-cognitive Computationalism

What I call *Eco-Cognitive Computationalism* sees computation as active in physical entities suitably transformed so that they become what I call *cognitive mediators* in which data can be encoded and decoded to obtain fruitful results. We have to immediately say that eco-cognitive computationalism, even if one sees cognition as computational, does not reduce computation to digital computation (that is to the processing of strings of digits according to rules).[1] When physical computation is seen in the perspective of the ecology of cognition it is easy, first of all, to understand the relationships between information, cognition, and computation taking advantage of the dynamic concepts of salience and pregnance derived from Thom's catastrophe theory,[2] but also the deep meaning of Turing's original ideas concerning the emergence of information, cognition, and computation in organic, inorganic, and artefactual agents, I will illustrate in this chapter. I hope it will become clear that eco-cognitive computationalism does not aim at furnishing an ultimate and static definition of the concepts of information, cognition, and computation, such as a textbook could provide, instead it intends, by respecting their historical and dynamical character, to propose an intellectual framework that depicts how we can understand their forms of "emergence" and the modification of their meanings.

Some preliminary explanations concerning Turing's concepts I will illustrate below in this chapter have to be provided. First of all Turing's speculations on how the so-called "unorganized brains"[3] are transformed in organized "machineries" are very important. Turing says that brains are continuous systems that can be treated as discrete systems able to perform "discrete" computations, so that we can describe the possible states of these brains as discrete sets, with the motion occurring by jumping from one state to another. Turing adds that all machines can be also considered as continuous (where the states constitute a continuous manifold and the behavior of the machine is depicted by a curve on this manifold). He also calls controlling machines the ones that only deal with information and active the machines that aim at generating some macroscopic physical effects: a bulldozer is a continuous and active machine, a telephone continuous and controlling. Brains, Turing says, can be conceived as machines and they are "probably"—continuous and controlling but "very similar to much discrete machinery" (Turing 1969 p. 5).

[1] It is well-known that this view is instead consistent with the classical view of cognition as the manipulation of linguistic/sentential entities.

[2] Catastrophe theory is a branch of mathematics that uses topology to explain events (such as a volcano eruption or an economical crisis) characterized by very sudden changes in behavior arising from relatively small changes in circumstances. It is part of bifurcation theory in the area of research on dynamical systems.

[3] More details about this concept proposed by Turing are illustrated below, at the end of this section.

In Turing's perspective we have to observe that brains are complex systems (organs) with many arbitrary multi-scale levels of description and representation. To avoid ambiguities in my use of Turing's concepts of discreteness and continuity it is better to explain the meaning in this context of the adjective "continuous": Turing himself understands that the nervous system is surely not a DSM (Discrete-State Machine) that is, in modern terminology, the brain rather could be better described as a kind of dynamical system that is sensitive to initial or limit conditions, more complex than any n-body physical system or turbulent stream. Turing calls a systems such as this one "continuous" both in his 1950 article (Turing 1950) and in his 1952 (Turing 1952) article on morphogenesis (Longo 2009a p. 380).[4]

Moreover, to also better clarify the concept of discreteness, we have to say that since Turing computer science suggests a perspective in which physical artefactual entities are "domesticated" to process the organization of information and knowledge into little boxes, bits, and pixels, which present discrete precision (each datum is well separated and accessible and each measure exact, in contrast to what happens for all—classical—physical processes), exactly, with no fuzziness and no contingency. We know that in chaotic deterministic systems, a fluctuation/variation below the interval of measurement induces radically different evolutions for the system. This, Turing observes in the 50s, is theoretically avoidable in the Discrete State Machine (DSM) he has invented and as he named it in those years (and such is also the case in practice). We also know that Turing machine is an abstract machine, a logical one, as he called it, with a Laplacian behavior, as a DSM the machine processes values which are necessarily discrete. The concept of program and the mathematical structure of its implementation are deterministic in Laplace's sense: the use of rules (or in the case of Laplacian mechanics of equations) entails predictability. Of course it is amazing that humans were able to transfer Turing's abstract machine, as a Discrete-State Machine (DSM) to physical machines, building artificial technological astonishing processes that are capable to evolve by discrete states, thanks to the role played by silicon electronic devices.[5]

---

[4] Longo further observes: "Note that we are not claiming here that the brain is a dynamical system: also Turing refers to the brain as, at least, a dynamical, highly sensitive, system [...]. To stay within his image, take a turbulent system that is at the same time very stable and very unstable, very ordinate and very inordinate; insert it sandwich-style between different levels of organization that regulate it and that it integrates. You will then have a very pale physical image of a biological entity. Among these entities, quite material, soulless and without software distinct from the hardware (the modern dualism of the cognitivism of the formal rule and of the program), you will also find bodies with nervous systems that integrate and regulate them (as networks of exchange and communication), within which they integrate themselves (as organs) and by which they are regulated (by hormonal cascades, for example)" (Longo 2009a, footnote 13, p. 390).

[5] "Of course, there may be some endowed indeterminacy (the machine can make steps which lead to an arbitrary element of a finite set of possible discrete states, instead of leading to a single one—we are then dealing with an non-deterministic DSM), but it consists of probabilistic type of abstract indeterminacy, already well studied by Laplace, and which is not the same mathematical concept as the unpredictability of deterministic dynamical systems, in the modern sense" (Longo 2009, p. 380).

By making a strange but fruitful analogy, Turing clearly says: "The cortex of an infant is an unorganized machinery, which can be organized by suitable interference training. The organization might result in the modification of the machine into a universal machine or something like it. [...] This picture of the cortex as an unorganized machinery is very satisfactory from the point of view of evolution and genetics" (Turing 1969 p. 16). I have to add that the analogy adopted by Turing "the cortex of an infant is an unorganized machine" is largely speculative and just plays a central heuristic role in his creative cognitive processes: obviously we know from basic neuroscientific research that brain is a highly organized system. Turing just refers to the fact that the cortices of infants are kinds of blank slates which are "socially" fulfilled through language. In this perspective the hypothesized random construction of the infant cortex does not deal with neurobiological aspects but to the basic absence of training, that is of signs coming from the external (also social) world through the senses—for example language. So the definition of "unorganized machinery" is related to a special level of description: Turing himself says that "It is conceivable that the same machine might be regarded by one man as organized and by another as unorganized" (Turing 1969 p. 9), see below Sect. 1.3.2.

This intellectual perspective first of all clearly depicts the evolutionary emergence of information, meaning, and of the first rudimentary forms of cognition, as the result of a complex eco-cognitive interplay and simultaneous *coevolution*, in time, of the states of brain/mind, body, and external environment. At the same time it furnishes the conceptual framework able to show how, thanks to an imitation of the above process, the subsequent invention of both the Universal Logical Computing Machine and the Universal Practical Computing Machine is achieved. In this last case, the dual invention is realized, as the externalization of computational capacities in those artefactual physical entities that compute for some human or artefactual agents: those computers that in this perspective offered by Turing I called "mimetic minds" (Magnani 2006) (the concept of mimetic mind is explained below in Sect. 1.3.1).

1.1.2 Computation and Digital Computation

I have said in the previous subsection that *eco-cognitive computationalism* sees computation as active in physical entities suitably transformed so that they become what I call *cognitive mediators*, in which data can be encoded and decoded to obtain fruitful results. I also said that eco-cognitive computationalism, even if sees cognition as computational, does not reduce computation to digital computation.

I have just illustrated in the previous subsection that I favor a perspective that sees the evolutionary emergence of information, meaning, and of the first rudimentary forms of cognition, as the result of a complex eco-cognitive interplay and simultaneous coevolution, in time, of the states of brain/mind, body, and external environment. This perspective excludes every long stability of the meaning of concepts such as computation, information processing, and cognition: for this reason it is important to avoid the identification of "computation" with digital computation (as implicitly delineated by the classical mathematical theory). We perfectly know that digital computation is obviously defined by classical theory of computation, which derives from Turing's and other researchers' discoveries and it is actually executed through digits in physical systems of various kinds (mechanical, electromechanical, electronic, magnetic, etc.): it involves algorithmic computation that can be performed by a Turing machine. We have also to note that in the last decades the notion of digital computation certainly played the role of modeling neural activity and of central features of human cognition, but many neuroscientists (for example (Port and Gelder 1995)) strongly contended that neural networks do not perform digital computation at all: in this last case cognition is trivially seen as performed by neural networks but is not computation.

In summary, the concept of computation I am adopting here goes beyond digital computation, that is the processing of strings of digits according to rules; it also refers to other computations performed by physical devices, such as for example brains, with their activity of modifications of the configurations of neural networks (and the electrical activity—neurons fire and provide electrical inputs to other neurons) and chemical processes (neurotransmitters and hormones), but also analog computers, and the so-called neural computers. It is an extended idea of the concept of computation: as I have already stressed above the concept must not be seen as static, it evolves depending on the transformations of theory and practice, as we will see in the following sections.[6]

1.1.3 Information Processing, Cognition, and Computation

Also the concept of information processing is subjected to changes of meaning, depending on the contexts and on the intellectual perspectives adopted. However, we can certainly say that information processing refers to the transformation of all kinds of information, from the perspective of the transmission of pure *signals* (or *signs*, if we want to adopt the Peircean terms), from those signs intended as subject to the constraints of communication theory (Shannon information, in which measures of information "are not" also measures of meaning), to the signs that carry an information accompanied by *meaning*. When information is also perceived by some agents as endowed with cognitive value, the problem of its emergence (and so of its generation, for example through abduction)[7] and of the role of *representations*

[6] On the related debate concerning the conflicting recent views on the relationships between computation and representation in cognitive neuroscience (and on computationalism about the brain, that is the view that the brain literally performs computations) cf. a recent special issue of the journal *Minds an Machines* (Piccinini 2018). I think this rich debate, mainly aiming at philosophical/ontological and definitory ambitions, still shows how the meaning of the concepts of representation and computation evolves depending on theory and practice.

[7] Abduction is a cognitive process that leads to hypotheses, both diagnostic—such as in medical reasoning—and creative, such as in the case of scientific discovery. I have richly illustrated the concept of abduction in some of my recent studies (Magnani 2009, 2017).

is fundamental. Of course, in the case of information as perceived by some agents as endowed with a cognitive value, falsity and truth are at play. Just to make an example, when in the case of medical diagnosis we have performed an abductive hypothesis and we consider the related information reached about the disease at play highly plausible (and able to permit us to *know* something relevant about the patient), higher can be considered the quantity of information channeled by the symptoms presented by a patient, on the basis of which the abductive process was performed. In sum, when information is also intended as a part of cognition, this refers to the fact that signs at stake have to be more or less meaningful.

When people ask "Is cognition computation?", we can simply answer, yes, it is, but it is not only computation, so that we cannot identify cognition with computation. Both information processing and computation are certainly involved in cognition, and to clarify what kinds of computation and information processing are involved in cognition is a respectable and difficult task,[8] even if it is destined to become outdated: I indeed contend that, given the fact the concept of computation evolves and changes, the other two concepts, already characterized in themselves by unstable definitions, subjected to the developments of knowledge, technologies, and cultural frameworks, also evolve.

Let us repeat, certainly cognition involves the transmission of signals (or signs in Peircean terms), which are subject to the constraints of communication theory (Shannon information), but cognition also refers to the fact that the signs themselves have to be meaningful. Furthermore, in the case of "cognition" signs are also composed by models, icons, etc., and not only by linguistic, or sentence-like structures, similar to the ones of natural language, such as it was maintained by the classical paradigm of symbolic AI. In cognition signs are also usually considered representational, so that cognition, when performed for example through digital computation, is carried out over representations. Connectionists[9] often contrasted the idea that cognition is based on a language of thought, favoring models deriving from the functions of the brain: we have to say that, presently, connectionists and computational neuroscientists simply contend that cognition is explained by neural network activity (Piccinini and Scarantino 2011 p. 15).

[8] Fresco (2013) fruitfully aims at disambiguating the concept of digital computation in contemporary cognitive science by illustrating how digital computation is implemented in physical systems, without ending up in pancomputationalism, that is, the view that every physical system is a digital computing system and can be explained in computational terms.

[9] Even if we have to remember that some connectionists considered cognition in terms of neural networks that perform computations defined over strings of digits.

1.2 The Emergence of Information and Cognition

1.2.1 Pregnances and Saliences as Information

I think the concept of pregnance, advanced by Thom's catastrophe theory (Thom 1972, 1980) and derived from Wertheimer's Gestaltic concept of *Prägnanz*, in the framework of the so-called "semiophysical" perspective, can be extremely useful to present an integrated illustration of the emergence of information (and consequently, of cognition). What is important here is to describe that pregnances can be created in situations of learning. Thom says that phenomenal discontinuities are captured by organisms as *salient forms* (such as in the case of a sound which interrupts the silence), so that the sensorial apparatuses echo in the organisms what happens outside in the physical environment.[10] After all, many components of organisms' neurophysiology are involved in the detection of the external (distal) environment and, possibly, in the production of an appropriate related deeper cognition.

The animal sensorial apparatuses obviously are collectors of "information" of various kinds, visual, olfactory, etc.: the case of the "cognitive" role of information embedded in "salient" forms can be (1) first of all seen in the case of the emergence of "special" information, which is endowed of some hardwired biological value. Moreover, when the sensorial activities are further specialized thanks to (2) not merely instinctual cognitive endowments, but thanks to plastic cognitive capacities, due to learning, language, and concepts, the information that can be gathered surely increases, and so its cognitive value. Indeed, it is clear that pregnances have both genetical and extra-genetical characters. In this last case they are transmitted in animals and human animals as "cognitive" packets of *learned knowledge*. The following example is instructive, a skilled botanist gathers more information through her sensorial apparatuses because she is reinforced by knowledge:

> Imagine a skilled botanist accompanied by someone like myself who is largely ignorant of botany taking part in a field trip into the Australian bush, with the objective of collecting observable facts about the native flora. It is undoubtedly the case that the botanist will be capable of collecting facts that are far more numerous and discerning than those I am able to observe and formulate, and the reason is clear. The botanist has a more elaborate conceptual scheme to exploit than myself, and that is because he or she knows more botany than I do. A knowledge of botany is a prerequisite for the formulation of the observation statements that might constitute its factual basis (Chalmers 1999 pp. 11–12).

In this example also plastic cognition (scientific) is at play, and the chances of collecting rich information are extremely augmented. This fact is also confirmed by recent research in cognitive science: thanks to an interdisciplinary approach and an appropriate experimentation some cognitive scientists (cf. for example (Raftopoulos 2001a, b)) have recently acknowledged the fact that, in humans, perception (at least in the visual case) is *semi-encapsulated*, that is it is not strictly modular, as Fodor (1984)

[10] An analogous perspective has recently been illustrated by Raja (2018) exploiting the well-known Gibson's ecological psychology (Gibson 1979): the animal perceives its environment by "resonating" to information in energy patterns.

argued, and so not encapsulated, hardwired, and domain-specific, neither wholly "penetrable" by higher cognitive states (such as desires, beliefs, expectations, etc.) Consequently, there is a considerable amount of information in visual perception which is, so to speak, theory-neutral, but also a notable certain degree of theory-ladenness is at work for example in the case of so-called "perceptual learning". The basic assumption regarding the cognitive impenetrability of perception is not put in question, but it is recognized that perception is informationally "semi-encapsulated" and so semi-hardwired, notwithstanding its main bottom-down character it is not insulated—so to speak—from "knowledge".[11]

By the way, as the reader can easily imagine, a perception experienced as salient is already fundamental to favor the emergence of an "information" that refers to individual forms and, consequently, also to delineate those entities which are at the basis of what we call "concepts": indeed when a human being, definitely, perceives a salience she is facing a kind of proto-concept to be here intended as a class of equivalence between forms which refer to the same case. Later on also a correspondent sign of every kind can be created, a word for example, or an icon, which refers to that recognized event: the process that Peirce called semiosis takes place, where signs as referring to their objects are at play. Following Peirce, we can say that human brains make up a series of signs that are used in building or showing or responding to a series of signs: being engaged in this activity, which is semiotic but at the same time patently cognitive they become "minds" able to think intelligently. Signs can be externalized in both natural and artefactual environments, for example in drawing and writing, giving rise to that process of externalization of the mind (or disembodiment of the mind), which is at the origin of the dominant distributed character of human cognition, as I will illustrate in Sect. 1.2.4. The emergence of drawing and writing is certainly related, as Longo (2009) maintains, to the human evolution to alphabetic natural language.

1.2.2 Pregnances and Paniformationalism

Pregnances, phenomena that can be either biological or physical, are in turn, Thom says, "non-localized entities emitted and received by salient forms" (Thom 1988 p. 16). Two clear examples can be given. The first one refers to natural phenomena: (1) infection is a pregnance (mediated by a virus, that is a material/biological medium) that affects healthy subjects, who are the invested saliences that in turn can re-emit that same contagion as a pregnance into the natural niche (in which, in turn, other media such as air or blood are the transmitters).

---

[11] Further insight on the semi-encapsulated character of cognition can be found taking advantage of the analysis of the several cultural forces that also transform cognitive actions, as illustrated by Nisbett (2003).

In this case it seems weird to contend that information is at play, even if in the literature there is even an all-encompassing notion of information, a kind of paninformationalism, in which physical (or biological) information is extrapolated to every state of a physical (or biological) system which is delineated as an information-carrying state (Wolfram 2002; Lloyd 2006). I am not inclined to follow this perspective in the case of the analysis given in this article, because I prefer to see information always referred to a biological or artefactual agent that "selects" it as such in the multiplicity of environmental events. Of course I am not undervaluing here the many excellent results that physicists (for example (Chiribella et al. 2012; Goyal 2011) (and logicians) have reached, providing mathematical frameworks for seeing quantum theory in the perspective of the principles of information processing. My only warning does not concern these results but the possible abuse of the notions of information (and of computation) in physics and biology, a problem extensively illustrated in detail in (Longo 2009, a).

The second example is related to not merely physical but also clear "cognitive" activities, in the case of some animals: (2) worker honeybees communicate with each other thanks to the signs embedded in the iconic movements expressed by a dance, which represent the pregnance that indicates the site where the insects have found food in order to "inform" the other cospecific individuals—that is the invested salience—about the location. Here the medium of the pregnance is the air that carries/transmits the ondulatory sounds and the light signals to the aim of generating a neurobiological effect at the destination, that is, something similar to what we humans call a "psychic" event.

In this second case we can contend that some transmission of *information* is at play, but also of *cognition*: after all, in animals the existence of a representation-oriented behavior can be postulated (Dummett 1993) where nonlinguistic pseudothoughts drive a plastic model-based "cognitive" activity. In this case the word cognition refers to processes that are not necessarily hard-wired but also the fruit of learning and productive activities, appropriate to certain situations, that animals—from mammals to bacteria—precisely learn and produce, such as in the case of the dance of the honeybees.

Obviously, also salient forms rooted in instinctual wired behavior channel "biological significance" (and consequently can be considered carriers of cognition), for example a prey for the aggressive predator, or the predator for its prey: *salient forms of this type are called pregnant.* In terms of abductive cognition we can say that a pregnant input is cognitively *highly diagnostic* and a trigger to start abductive production of hypotheses: Charles Sanders Peirce already observed that in the case of a chicken which perceives the hardwired pregnance of food the instinctual hypothesis concerning its eatability is immediately formed.

1.2.3 The Emergence of Information

At this point a solution to the conundrum of the emergence of information and cognition can be found, taking advantage of our evolutionary and naturalistic perspective, influenced by the semiophysical perspective derived from Thom's catastrophe theory. It is important to analyze this kind of solution because we will have to compare it to the process of the emergence of computation, I will illustrate below in Sect. 1.3. Not only, my view of the intertwining between information, cognition, and computation will become clear when in Sect. 1.3.1 I will delineate the concept of *mimetic mind*. Pregnances, and the information and cognition they embed, can be abductively activated, as I have illustrated in previous subsections, but also created for the first time.

A simple example is of help: when bell ringing is reiterated often enough together with the exposition of a slice of meat to a dog, Pavlov's conditioning process explains that the nutritive pregnance of meat is disseminated in the environment by contiguousness to the salient auditory form. What happens? The salient form, in this case the sound of the bell, which is also an *information* spread in the environment, is invested with the pregnance of the meat. Now the *cognitive* endowments of the dog are enriched by a new aspect: the "information" carried by the bell (referring to the possibility of eating) now is endowed with a new *created* meaning (related to a sound, we can of course consider a *sign*, in semiotic terms) for that animal agent. The information, already salient before, now is also pregnant, full of meaning. Thom adopts the exoteric image of an invasive fluid, it is useful to quote: "So we can look on a pregnance as an invasive fluid spreading through the field of perceived salient forms, the salient form acting as a 'fissure' in reality through which seeps the infiltrating fluid of pregnance" (Thom 1988 p. 7).[12] Of course the pregnance can be reinforced through repetition and/or similarity and step after step the bell—in the new formed cognitive features belonging to the "psychism" of the dog—will refer solidly to the meat. Of course the cognitive aspects are extremely dynamic because they are related to pregnances that can also vanish when reinforcement is no more possible, or distant from the organism at play, or when, given this absence, the salient form is interested by another (eventually new) pregnant form.

1.2.4 External Artifacts, External Representations and the Emergence of Cognition

I have said above (Sect. 1.2.1) that signs can be externalized in both natural and artefactual environment, for example in drawing and writing, giving rise to that process of externalization of the mind (or disembodiment if the mind), which is at the

[12] To describe the emergence of pregnances Thom adopts the Pavlovian perspective. Other perspectives take advantage of Hebbian (1949) and other more adequate learning principles and models, cf. for example (Loula et al. 2010).

origins of the dominant distributed character of human cognition. In Sect. 1.3 I will show that Turing is perfectly aware of the fact (Turing 1969) that a big cortex can furnish an evolutionary advantage when is accompanied by a great quantity of relevant information and knowledge already stored in external props; only sufficiently evolved human collectives can have at their disposal this storage, such as data from paleoanthropology seems to teach (Mithen 1996, 1999). Indeed, storing in the external environment signs and drawings, and manipulating external entities transforming them into artifacts, is the main process that characterizes the birth of the so-called material culture (related to the construction of the first handaxes, about 1.4 million year ago, the birth of what Mithen also classifies the "technical intelligence" of the primitive human mind). It is important to stress that, already at these early stages of human evolution, cognition is distributed, at the same time it is a process of delegation of "information" to external tools, props, and devices and the starting of an activity of rich semiosis. Not only, Mithen observes that "The clever trick that humans learnt was to disembody their minds into the material world around them: a linguistic utterance might be considered as a disembodied thought. But such utterances last just for a few seconds. Material culture endures" (Mithen 1999 p. 291). We can conclude saying, according to recent intellectual research in cognitive science (Clark 2008), that minds are "extended" and artificial in themselves. We can say that in a certain sense, through the processes I have just sketched, human primitive brains become a sort of universal "intelligent" machines endowed with a rich activity of consciousness.

It is also important to stress that the reference to the role of external representations and to the disembodiment of the mind is able to account in a very satisfactory way central dynamic aspects of human cognitive processes. So to speak, brains/minds "extend" themselves in the external world to the aim of semiotically representing signs, words, icons, drawings in rocks, boards, sheets of paper, materials of various types, also to the aim of using them as anchors for helping to generate new thoughts, ideas, and concepts that do not have a "natural home" internally, within them. New information and new knowledge are usually produced inside this distributed process of semiotic activity.

The reader that is further interested in this activity of emergence and subsequent growth of information and knowledge in this distributed perspective and with reference to hypothetical reasoning can see chapters two and three of my book (Magnani 2009). The role of external representations and of the related delegations of cognition to external tools and artifacts is illustrated taking advantage of examples coming not only from the paleoanthropological studies concerning the cognitive processes of the primitive mind but also in the case of mathematical and scientific knowledge.

1.3 The Emergence of Computation

1.3.1 Turing on the Emergence of Information, Cognition, and Computation in Organic, Inorganic, and Artefactual Agents

The birth of computation is strictly related—in Turing's seminal work *Intelligent Machinery* (1948)—to heuristic cognitive processes that take advantage of the study of the role in the external environment of organic bodies, physical entities, and artifacts.

Reflecting upon the possibility of constructing intelligent machines, Turing immediately says that human intelligence can be produced only if a suitable education is performed (Turing 1969 p. 3) and indicates an analogy between human brains (that is organic entities) and computational machines (that is physical artefactual entities). He proposes the concept of *unorganized machine* and provides the example of the infant human cortex, which is a "natural" organic entity that can be educated by means of "rewards and punishments". Together with unorganized machines Turing also lists *paper machines* (see below the following subsection) and the two famous new fundamental kinds of machineries he invents: the *(Universal) Logical Computing Machines* (LCMs), which are considered discrete machines, and the *(Universal) Practical Computing Machines* (PCMs), which are machines in so far as they are external physical entities (inorganic and "artefactual") that manage their saved information in a form very dissimilar from the tape form, but that reflect the fact that "[......] given any job which could have been done on an LCM one can also do it on one of these digital computers" (Turing 1969 p. 8).

LCM is a kind of discrete machine Turing introduced in 1937 that has

> [...] an infinite memory capacity obtained in the form of an infinite tape marked out into squares on each of which a symbol could be printed. At any moment there is one symbol in the machine; it is called the scanned symbol. The machine can alter the scanned symbol and its behavior is in part described by that symbol, but the symbols on the tape elsewhere do not affect the behavior of the machine. However, the tape can be moved back and forth through the machine, this being one of the elementary operations of the machine. Any symbol on the tape may therefore eventually have innings (Turing 1992 p. 6).

PCMs are machines that organize their stored information in a form that does not repeat the tape form. In presence of a huge quantity of steps that can be involved in computation along with the tape PCMs machines can easily solve the problem because, Turing says "[...] by means of a system that is reminiscent of a telephone exchange it is made possible to obtain a piece of information almost immediately by 'dialing' the position of this information in the store" (p. 8). Turing also observes, as already said, that "nearly" all the PCMs under construction have the fundamental properties of the Universal Logical Computing Machines: "[...] given any job which could have been done on an LCM one can also do it on one of these digital computers" (*ibid.*), it is in this perspective that we can speak of Universal Practical computing Machines.

I contend that the interest of Turing's discoveries and observations is both historical and epistemological. In the quoted seminal article he first of all considers unorganized and other kinds of various supports as physical entities of computation, which is consequently regarded as active in physical (of course also in biological) entities suitably modified, thanks to "education", seen as a certain set of transformations. Programming, Turing says, mimics education in the hope that the machine will be capable of presenting certain expected reactions to particular commands.

I have illustrated in the previous section that research in distributed cognition contends that humans delegate cognitive (and epistemic, moral, etc.) roles to externalities (artifacts, props, etc.) and then tend to recapitulate what they have checked occurring outside, over there in the external representations, after having manipulated—often with creative outcomes—the external invented structured model. For example, we know that it is not neurologically easy to perform a trivial multiplication of numbers by projecting in our mind the standard schema of that multiplication as it is used, with paper and pencil, taking advantage for example of a blackboard; to solve the small problem we instead need the manipulation of an external device, for example a sheet of paper. Mind puts representations in the environment, so reifying itself in various semiotic structures that are able not only to mimic but also to improve and modify its internal representations.

In the perspective of distributed cognition we can say that the (Universal) Logical Computing Machine (LCM) is added by Turing to the external idealized tools already objectified during the history of western civilization: it is a new abstract tool/machine that is characterized by very powerful mimetic properties, it is able to mimic plenty of human cognitive operations, and at the same time clarifies the concept of effective procedure, it is a real *mimetic mind*. On the basis of this abstract LCM machine Turing, already in 1950, contends that digital computers (as external physical suitable entities) can be constructed, and can mimic the actions of a human computer very closely. To express this action of externalization of an abstract and of a concrete tool performed by Turing's discoveries I called (Magnani 2006, 2009, Chap. 3) the (Universal) Logical Computing Machine (LCM) (the theoretical artifact) and the (Universal) Practical Computing Machine (PCM) (the practical artifact) *mimetic minds*: indeed they are able to mime the mind in a real universal way and because they are "minds" externally objectified in the environment, as distinctly describable formal intellectual structures and practical machines. We do not have to refer to a plenty of several machines that do a variety of jobs: it is just necessary to "program" this universal machine to have those jobs done. I think these mimetic minds represent a kind of triumph of that process of externalization of cognitive powers to the external environment at work since our ancestors times I have illustrated in the previous section.

1.3.2 Unorganized and Paper Machines

The heuristic process that leads Turing to the invention of both LCMs and PCMs is extremely interesting. Turing says that there are some entities, he calls *unorganized machines*, which are basically random "in their construction":

> So far we have been considering machines which are designed for a definite purpose (though the universal machines are in a sense an exception). We might instead consider what happens when we make up a machine in a comparatively unsystematic way from some kind of standard components. [...] Machines which are largely random in their construction in this way will be called "Unorganized Machines". This does not pretend to be an accurate term. It is conceivable that the same machine might be regarded by one man as organized and by another as unorganized (Turing 1969 p. 9).

Many units form machines of such a type. Each unit is characterized by two input terminals and has an output terminal that can be linked to the input terminals of 0 or more of other units. Turing gives an example of what he calls unorganized A-type machine, in which all units connected to a synchronizing unit from which synchronizing pulses are emitted at more or less equal intervals of times is given in Fig. 1.1 (Turing calls the times when the pulses arrive "moments" and each unit is capable of having two states at each moment). A-type unorganized machines, Turing adds, are a good and simple model of organic entities such as a nervous system with a *random arrangement of neurons*.[13]

Furthermore, "It is possible to produce the effect of a computing machine by writing down a set of rules of procedure and asking a man to carry them out. [...] A man provided with paper, pencil and rubber, and subject to strict discipline, is in effect a universal machine" (p. 9). Turing calls this kind of machine "Paper Machine".

It is more than evident than in both cases high level cognitive activities are anticipatorily described as dynamical activities (Giunti and Pinna 2016), which involve the coordinate intervention of both internal (mental) resources and external material ones, like paper and pencil and other suitable entities, in a situation of agent-environment interaction. Recent approaches in terms of dynamic systems (Port and Gelder 1995), as I have already stressed, in terms of distributed cognition (Clark 2008), can be illuminating in interpreting these extraordinary Turing's ideas. I have to point out that also Turing's invention of the LCM machine is performed thanks to a heuristic which takes advantage of a clear attention to the interplay between internal and external resources: when Turing stressed the active role in computation of what he calls "mimicking education", I will describe in the following subsection, he strongly pays attention to the central role played by the flux of information that comes from the external environment. The reader perfectly knows that the LCM standardly incarnates the view of the so-called classic computationalism, which refers to the idea that computations are performed by a cognitive system through internal symbolic transformations depending on purely syntactic rules: we are instead seeing that the emergence of this concept is in Turing completely related to a perspective that takes advantage of the consideration of the interaction between internal and

[13] On the infant cortex as an unorganized machine see Sect. 1.3.3.

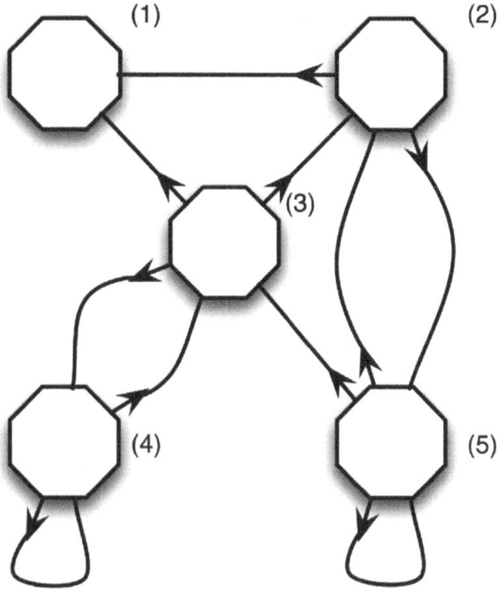

Fig. 1.1 An example of the so-called unorganized A-type machine, a simple model of a nervous system with a random arrangement of neurons

external aspects such as organic agents, paper and pencil, physical electronic artifacts, education, etc. These are not accessory to the new idea of computation, but an essential part of it. After all, just to make a final simple example, how can we grasp Turing's concept of infinite tape without referring to a finite material tape in an external environment?

1.3.3 The Emergence of Computation in Digital Physical Entities Seen as Mimicking Human Education

The process of education of human beings can be usefully exploited to provide a model of what Turing calls "education of machinery". The reader does not have to misunderstand the actual meaning of this analogy. Presently, the computer training can be certainly seen as quite different from human education, for example because in the latter case the semantic information plays the central role whereas in the former case, the semantic information plays no role. It is instead important to note that the analogy adopted by Turing performs a dominant heuristic role in his discovery process to the new concept of universal logical machine: in this case the few and simple similarities are important is so far as they produce new insights, and in this perspective the differences between the two cases are obviously not taken into account.

Indeed, by "mimicking education, we should hope to modify the machine until it could be relied on to produce definite reactions to certain commands" (Turing 1969 p. 14). The same happened in the case of a graduate student that took advantage of interactions (interferences we can say) with other human beings for a couple of decades so that "[…] a large number of standard routines will have been superimposed on the original pattern of his brain" (*ibid.*) Turing also observes that (1) those interferences are mainly with other men thanks to visual and other kinds of stimuli; (2) a man in concentrating simulates a machine without interference even if is conditioned by previous interactions.

When an unorganized machine (such as an infant cortex) is subjected to appropriate interferences it starts to behave as an organized (and possibly universal) machine to reach a determinate target. We have to note that in such cases the machines reach stable configurations as states of discrete machineries. The machines described in these sections are *discrete* machines because their states can be depicted as discrete sets, with the movements of the machines as activities of jumping from one state to another.

An example of a B-type unorganized machine is provided: if units are sufficiently numerous and we give it suitable initial conditions, the machine is transformed into a universal machine also able to become a repository of information. The establishment of the appropriate initial conditions has to be seen as a method for "organizing the machine", as a "modification" of an unorganized machine, as I have illustrated above.

I have anticipated that, following Turing, the infant brain can be conceived as an unorganized machine: "The cortex of an infant is an unorganized machinery, which can be organized by suitable interference training. The organization might result in the modification of the machine into a universal machine or something like it. […] This picture of the cortex as an unorganized machinery is very satisfactory from the point of view of evolution and genetics" (Turing 1969 p. 16). Not only, in this evolutionary perspective, which is certainly speculative but anticipatory of contemporary results of cognitive paleanthropology,[14] the existence of the human cortex can be justified only in terms of its organization, in a kind of *coevolution* between human cortex and external information available to organize it: "[…] the possession of a human cortex (say) would be virtually useless if no attempt was made to organize it. Thus if a wolf by a mutation acquired a human cortex there is little reason to believe that he would have any selective advantage" (*ibid.*) Consequently, the use of a big cortex (that is its suitable organization) demands an appropriate environment: "If however the mutation occurred in a milieu where speech had developed (parrot-like wolves), and if the mutation by chance had well permeated a small community, then some selective advantage might be felt. It would then be possible to pass information on from generation to generation" (*ibid.*) Turing's argumentation about the "unorganized"

[14] On the related evidence from cognitive paleanthropology about the so-called "disembodiment of the mind" and on what I called semiotic brains, as brains that become capable to make up a series of signs and that are engaged in making or manifesting or reacting to a series of signs, see (Magnani 2009, Chap. 3).

brains, considered as kinds of blank slates which are socially fulfilled through language has to be remarked because, even if speculative and mainly endowed with a dominant heuristic role, as I have already noted, demonstrates unusual attention to the importance of phylogenetic mechanisms present in human cognition, attention much more extended with respect to the one adopted by the traditional philosophical western schools.

Speech and Social Background

Turing's image is powerful. To transform the unorganized human cortex in a universal machine it is necessary to have at our disposal that kind of *information* that is granted by *speech* (even if very simple as in the case of parrot-like wolves) and at the same time a *social background* in which various "techniques" are available and learnable:

> [...] the isolated man does not develop any intellectual power. It is necessary for him to be immersed in an environment of other men, whose techniques he absorbs during the first twenty years of his life. He may then perhaps do a little research of his own and make a very few discoveries which are passed on to other men. From this point of view the search for new techniques must be regarded as carried out by human community as a whole, rather than by individuals (Turing 1969 p. 23).

We can derive a clear consequence: a big cortex represents an evolutionary advantage only if it is fecundated by a great storage of *meaningful information* also endowed with a *cognitive* value (knowledge) carried by external supports and tools that only an already developed collective of humans can have.

Rewards and Punishments

Turing's argumentation is close to the target of demonstrating that programming a machine mimics education. To train a child rewards and punishments are needed and this fact indicates that organization can happen only through two inputs. At this point Turing presents the example of an unorganized P-type machine, considered as a LCM without a tape and mostly incompletely described, a machine that thanks to suitable stimuli of pleasure and pain (and the provision of an external memory) can be transformed in a universal machine (p. 20). One could ask: "given that Turing was wrong on some things—e.g. 'Brains of the human infant are what he calls unorganized machines that can be educated thanks to 'rewards and punishments'"'—does falsity affect negatively his argumentations? In this case the answer is, as I have already explained above, that we are facing speculations about phylogenetic/ontogentic aspects that just play a lucky heuristic role in his invention of the logical computing machine, and that this heuristic character is what counts, even if in the light of current scientific results the concept appears unsatisfactory.

Discipline and Initiative

Hence, the infant cortex is transmuted in an intelligent one thanks to discipline but also thanks to initiative, which Turing considers the two main aspects of a process that has to be studied as it happens in humans to "copy it in machines" (p. 21). A prototypical situation that commands initiative is for example "Find a number n such that ...", "see if you can find a way of calculating the function which will enable us to obtain the values for arguments ...", that is totally equipollent to that of constructing a *program* to put on the machine at stake.

It seems easy for Turing to exploit the idea of human education I have illustrated above to build the new concept of computation. As I have already noted the metaphor of human education adopted by Turing plays a useful heuristic role in the discovery process that leads to the new idea of universal logical computing machine. Of course, as already said, Turing's picture of human development is speculative and unsatisfactory in the light of recent cognitive studies.[15] We can consider Turing's speculations a first insight on what he calls "unorganized machineries", but researchers in psychology, at least starting from Piaget, have fulfilled empirically very abstract notions such as the Turing's ones.

Indeed Turing observes that universal machines are usually subjected to two types of interference: *screwdriver* interference is when components of the machine are eliminated and substituted with others, originating entirely new machines; *paper* interference is when the insertion of new information in the machine changes its behavior. Of course we have to stress the fact that paper interference furnishes information that is both external and material. Turing also considered it possible to build a thinking machine because human beings have been already able to imitate various parts of man (microphone, television, etc.) Also the functions of nerves can be copied thanks to suitable electrical models given the fact the electrical circuits of electronic computing machineries show a capacity to carry and store information analogue to the ones of nerves. In this case Turing obviously refers to information carried out by means of digital computation, and he at the same time always stresses that there are not digital ways of encoding information in organic agents.

As I have illustrated above, Turing contends that a big cortex represents an evolutionary advantage only if it is fecundated by a great storage of meaningful information and knowledge carried by external supports and tools that only an already developed collective of humans can have. In few words, *information*, but also *cognitive* contents expressed thanks to language, signs, icons, etc. have to be already present. Similarly Turing for example contends that transforming an external physical artefactual device (such as an electronic machine) *paper* interference is needed, when the insertion of new information in the machine changes its behavior. In this perspective the digital machine (a discrete state machine) is surely an alphabetic machine, made possible thanks to the human evolution to alphabetic natural language (of course, as we have noted, we know it is also a logical and formal machine, LCM) (Longo 2009). Longo thinks, and I agree with him, that this fact is at the root of that enormous "discretization of knowledge" that the birth of computation has generated. The "continuous" natural language is indeed transformed by the alphabet in something divided into undistinguished atoms, forming letters. They are in themselves without any kind of meaning, a meaning that instead emerges by combining them syntacti-

[15] I address the readers interested in the psychological aspects of evolving cognitive processes from child to adults to the rich handbook (Goswami 2010) and, more oriented to theoretical and cognitive aspects, to Karmiloff-Smith et al. (2018).

cally when competent human agents are able to sense and manipulate them. Given the discreteness of digital machines that is at the basis of their *imitation* power—they are mimetic machines, as I have already stressed—Turing contrasted this simple imitation power to the much stronger epistemological power of the *modeling* capacity of mathematics.

Summary

By exploiting the dynamic concepts of salience and pregnance, derived from Thom's catastrophe theory, I have illustrated the emergence in humans of information, meaning, and of the first kinds of cognition as the outcome of dynamic coevolutionary interactions between brain/mind internal processes, body itself, and external environment. Thanks to this analysis I proposed a clarification of the concepts of information, cognition, and computation stressing that their dynamic character can only be described taking into consideration their conceptual intertwining from both a cognitive and an epistemological perspective. In the second part of the chapter, seeing physical computation in the light of the ecology of cognition I offered a new analysis of Turing's original ideas concerning the emergence of information, cognition, and computation in organic, inorganic, and artefactual agents. I especially emphasized Turing's new ideas on the emergence of computation in digital physical entities as related to an analogy to the activity of mimicking human education. Considering computation in the perspective of the above cognitve/epistemological analysis and as dynamically active in distributed physical entities of various kinds, suitably transformed so that data can be encoded and decoded to obtain appropriate results, sheds first light on what I call "eco-cognitive computationalism".

References

Chalmers, A.F.: What is this Thing Called Science. Hackett, Indianapolis/Cambridge (1999)

Chiribella, G., D'Ariano, G.M., Perinotti, P.: Quantum theory, namely the pure and reversible theory of information. Entropy **14**, 1877–1893 (2012)

Clark, A.: Supersizing the Mind. Embodiment, Action, and Cognitive Extension. Oxford University Press, Oxford/New York (2008)

Dummett, M.: The Origins of Analytical Philosophy. Duckworth, London (1993)

Fodor, J.: Observation reconsidered. Philosophy of Science, 51, 23–43. Reprinted in Goldman (1993), 119–139 (1984)

Fresco, N.: Physical Computation and Cognitive Science. Springer, Cham, Switzerland (2013)

Gibson, J.J.: The Ecological Approach to Visual Perception. Houghton Mifflin, Boston, MA (1979)

Giunti, M., Pinna, S.: For a dynamical approach to human computation. Logic J. IGPL **24**(4), 557–569 (2016)

Goldman, A.I. (ed.): Readings in Philosophy and Cognitive Science. Cambridge University Press, Cambridge, MA (1993)

Goswami, U.: The Wiley-Blackwell Handbook of Childhood Cognitive Development. Wiley, Malden, MA (2010)

Goyal, P.: Information physics—towards a new conception of physical reality. Information **3**, 567–594 (2011)

Hebb, D.O.: The Organization of Behavior. Wiley, New York (1949)
Karmiloff-Smith, A., Thomas, M.S.C., Johnson, M.H. (eds.): Thinking Developmentally from Constructivism to Neuroconstructivism. Selected Works of Annette Karmiloff- Smith. Routledge, Oxford (2018)
Lloyd, S.: Programming the Universe: A Quantum Computer Scientist Takes on the Cosmos. Knopf, New York, NY (2006)
Longo, G.: Critique of computational reason in the natural sciences. In: Gelenbe, E., Kahane, J.-P. (eds.) Fundamental Concepts in Computer Science. Imperial College Press/World Scientific, London (2009a)
Longo, G.: Turing and the "imitation game" impossible geometry. Randomness, determinism and programs in Turing's test. In: Epstein, R., Roberts, G., Beber, G. (eds.) Parsing the Turing Test. Philosophical and Methodological Issues in the Quest for the Thinking Computer, pp. 377–411. Springer, Dordrecht (2009b)
Loula, A., Gudwin, R., El-Hani, C.N., Queiroz, J.: Emergence of self-organized symbol-based communication in artificial creatures. Cognitive Syst. Res. **11**(2), 131–147 (2010)
Magnani, L.: Mimetic minds. Meaning formation through epistemic mediators and external representations. In: Loula, A., Gudwin, R., Queiroz, J. (eds.) Artificial Cognition Systems, pp. 327–357. Idea Group Publishers, Hershey, PA (2006)
Magnani, L.: Abductive Cognition. The Epistemological and Eco-Cognitive Dimensions of Hypothetical Reasoning. Springer, Heidelberg/Berlin (2009)
Magnani, L.: The Abductive Structure of Scientific Creativity. An Essay on the Ecology of Cognition. Springer, Cham, Switzerland (2017)
Mithen, S.: The Prehistory of the Mind. A Search for the Origins of Art, Religion, and Science. Thames and Hudson, London (1996)
Mithen, S.: Handaxes and ice age carvings: Hard evidence for the evolution of consciousness. In: Hameroff, A.R., Kaszniak, A.W., Chalmers, D.J. (eds.) Toward a Science of Consciousness III. The Third Tucson Discussions and Debates, pp. 281–296. MIT Press, Cambridge (1999)
Nisbett, R.E.: The Geography of Thought: How Asians and Westerners Think Differently. and Why. Free Press, New York (2003)
Piccinini, G. (ed.).: Computation and representation in cognitive neuroscience. Special Issue of the J. Minds Mach. **28**(1) (2018)
Piccinini, G., Scarantino, A.: Information processing, computation, and cognition. J. Biol. Phys. **37**(1), 1–38 (2011)
Port, R.F., van Gelder, T. (eds.): Mind as Motion. Explorations in the Dynamics of Cognition. The MIT Press, Cambridge, MA (1995)
Raftopoulos, A.: Is perception informationally encapsulated? The issue of theory-ladenness of perception. Cognitive Sci. **25**, 423–451 (2001a)
Raftopoulos, A.: Reentrant pathways and the theory-ladenness of perception. Philos. Sci. **68**, S187–S189. Proceedings of PSA 2000 Biennal Meeting (2001b)
Raja, V.: A theory of resonance: Towards an ecological cognitive architecture. Minds Mach. **28**(1), 29–51 (2018)
Thom, R.: Stabilité Structurelle et Morphogénèse. Essai d'une théorie générale des modèles. Translated by D. H. Fowler, Structural Stability and Morphogenesis: An Outline of a General Theory of Models, W. A. Benjamin, Reading, MA, 1975. InterEditions, Paris (1972)
Thom, R.: Modèles mathématiques de la morphogénèse. Translated by W. M. Brookes and D. Rand, Mathematical Models of Morphogenesis, Ellis Horwood, Chichester, 1983. Christian Bourgois, Paris (1980)
Thom, R.: Esquisse d'une sémiophysique. Translated by V. Meyer, Semio Physics: A Sketch, Addison Wesley, Redwood City, CA, 1990. InterEditions, Paris (1988)
Turing, A.M.: Computing machinery and intelligence. Mind **49**, 433–460 (1950)
Turing, A.M.: The chemical basis of morphogenesis. Philos. Trans. Royal Soc. **237**(641), 37–72 (1952)

Turing, A.M.: Intelligent machinery [1948]. In: Meltzer, B., Michie, D. (eds.) Machine Intelligence, vol. 5, pp. 3–23. Edinburgh University Press, Edinburgh (1969)

Turing, A.M.: Collected Works of Alan Turing. Mechanical Intelligence. Ince, D.C. (ed.) Elsevier, Amsterdam (1992)

Wolfram, S.: A New Kind of Science. Wolfram Media, Champaign (2002)

Chapter 2
Eco-cognitive Computationalism

We have already delineated some basic aspects of the so-called eco-cognitive computationalism, for example the fact that computation is always seen in context, exploiting the ideas developed in those projects that have originated the recent views on embodied, situated, and distributed cognition. As illustrated in the previous chapter Turing's original intellectual perspective has already clearly depicted the evolutionary emergence in humans of information, meaning, and of the first rudimentary forms of cognition, as the result of a complex interplay and simultaneous coevolution, in time, of the states of brain/mind, body, and external environment. This cognitive process played a fundamental heuristic role in Turing's invention of the universal logical computing machine. It is by further extending this eco-cognitive perspective that we can see that the recent emphasis on the simplification of cognitive and motor tasks generated in organic agents by morphological aspects implies the construction of appropriate "mimetic bodies", able to render the accompanied computation simpler, according to a general appeal to the "simplexity" of animal embodied cognition. I hope it will become clear that, as already contended in the previous chapter, eco-cognitive computationalism does not aim at furnishing a final and fixed definition of the concept of computation, but stresses the historical and dynamical character of the concept.

2.1 Computationalism in an Eco-cognitive Perspective

As I have already illustrated in the previous chapter, what I call *Eco-Cognitive Computationalism* sees computation in context, exploiting the ideas developed in those projects that have originated the recent views on embodied, situated, and distributed cognition: computation is seen as active in physical entities appropriately transformed so that they become what I call *cognitive mediators*, in which data can be

L. Magnani, *Eco-Cognitive Computationalism*, Cognitive Systems Monographs 43,
https://doi.org/10.1007/978-3-030-81447-2_2

encoded and decoded to obtain productive results. This eco-cognitive approach to the concept of computation is strictly related to the tradition of research concerning the so-called distributed and embodied cognitive systems. The "distributed" approach sees cognition as a socially distributed process, pragmatically oriented, and affirms that cognitive processes can be better analyzed as situated in and distributed across material artifactual circumstances, in which the "ecological" view also emphasizes the role of the agent-environment interaction. The theory of distributed cognition was created by Hutchins (1995, 2010) to describe in a novel and more satisfactory way various problem solving processes in real work situations, so providing a new perspective that encountered great success in general cognitive science. I consider this approach particularly appropriate to treat the concept of computation: I will try to show that using this perspective both Turing's discoveries—examined in the previous chapter—and the recent new deal on morphological computation can be better understood and at the same time seen as sharing analogous epistemological characters. Indeed, eco-cognitive computationalism does not reduce computation to digital computation (that is to the processing of strings of digits according to rules), but it is open to include and consider other and new forms of computation. The perspective of the ecology of cognition concentrates on "physical computation", exactly following Turing's original thoughts concerning the emergence of computation in organic, inorganic, and artefactual agents.

Turing's germinal speculations on how the so-called "unorganized brains" and "unorganized machines" [1] are transmuted into organized "machineries" are extremely interesting: the problem is to see how, so to speak, "innocent" entities (in the sense of virginal objects "free from cognitive capacities") become first of all carriers of information and knowledge, but also carriers of computation. Turing is convinced that the emergence of rudimentary forms of information and cognition can be clearly comprehended in an evolutionary perspective, as the fruit of an eco-cognitive interplay and simultaneous *coevolution*, in time, of the states of brain/mind, body, and external environment. At the same time, this evolutionary perspective favors the subsequent creation of both the (Universal) Logical Computing Machine (which illustrates computation as a pure syntactic process) and the concrete (Universal) Practical Computing Machine.[2] Machines lose their cognitive innocence (they were already merely involved in elementary low-level cognitive tasks, such as in the case of telephone, telegraph, etc.) and become *universal* cognitive physical entities, in so far as they become *computational* artifacts that compute for humans or artefactual agents: those computers that in this perspective offered by Turing I called "mimetic minds" (Magnani 2006), as I have illustrated in the previous chapter, Sect. 1.3.1. I hope it will become clear that eco-cognitive computationalism does not aim at furnishing a final and stable definition of the concept of computation, such as an introductory treatment or a different approach could provide: I intend to take into account the

[1] On the meaning of these concepts in Turing's sense see the previous chapter and Sects. 2.2.1 and 2.5.

[2] Further details on this view proposed by Turing are illustrated in Magnani (2018).

historical and dynamical character of this concept, to propose an intellectual framework that depicts how we can understand not only the change of its meaning, but also the "emergence" of new forms of computations.

2.2 The Birth of *Mimetic Minds*: Educating Human Brains/Educating Physical Entities

I have said in the previous section that what I call *eco-cognitive computationalism* sees computation as active in physical entities suitably transformed so that they become what I call *cognitive mediators*, in which data can be encoded and decoded to obtain useful results. I also stressed the importance of an evolutionary framework that sees the emergence of information, cognition, and computation as the result of a complex eco-cognitive interaction and simultaneous coevolution, in time, of the states of brain/mind, body, and external environment. As I have stressed in the previous chapter, the concepts of information, cognition, and computation must not be considered as static, and their meaning changes depending on the modifications of theory and practice: in this chapter, I will present the new variation of the meaning of the concept of computation generated by the involvement of morphological aspects.

Following the Peircean semiotic perspective, we can remember that signs can be externalized in both natural and artefactual environments and we could add, using a term coming from the last decades of research in cognitive science, that signs are *distributed* and so externalized in a process also called "disembodiment of the mind". We have already illustrated that in his seminal lecture *Intelligent Machinery* (1948) (Turing 1969), Turing describes this same process adopting some neurological metaphors: a big cortex can present an evolutionary advantage when two conditions are fulfilled: (1) the presence of a huge quantity of relevant signs (information and knowledge) available in an environment suitably "artificialized" (props, supports, etc.); and (2) the embedment, discretely developed, in a small community able to manage information.[3] The birth of computation is indeed interestingly linked—in Turing's work I have just quoted—to creative heuristics intertwined with the analysis of the function—out there, in the external environment—of organic bodies, physical entities, and artifacts.

2.2.1 *Educating Human Brains*

Turing contends that human intelligence is obviously generated by an appropriate education (Turing 1969, p. 3) and an analogy between human brains as organic entities and computational machines as physical artifacts has to be built. Brains of

[3] In Magnani (2006), I have extendedly explained that this speculative argumentation has been recently supported by research in paleoanthropology on the birth of material culture.

the human infant are similar to what he calls *unorganized* machines—machines which are largely random in their construction—that can be educated thanks to "rewards and punishments". The analogy created by Turing "the cortex of an infant is an unorganized machine" (Turing 1969, p. 16) is speculative and just plays a central heuristic role in his creative cognitive processes: obviously, we know from basic neuroscientific research that the brain is a highly organized system. Turing just refers to the fact that the cortices of infants are kinds of blank slates which are "socially" fulfilled through language. From this perspective, the hypothesized random construction of the infant cortex does not deal with neurobiological aspects but with the absence of "education" coming from the external social world—for example, language. Thus, the definition of "unorganized machinery" is related to a special level of description.

When an unorganized machine (such as an infant cortex) is submitted to suitable interferences, its behavior changes, and step after step the machine becomes organized (and possibly also universal). Not only, existence of a human cortex is evolutionarily justifiable solely in terms of its organization in the framework of a *coevolution* between it and the external information available to organize it: "[...] the possession of a human cortex (say) would be virtually useless if no attempt was made to organize it. Thus if a wolf by a mutation acquired a human cortex there is little reason to believe that he would have any selective advantage" (Turing 1969, p. 16). An environment full of information (made real thanks to speech and at the same embedded in a *social background* in which several "techniques" are usable and learnable) has to be already available, but also, as I have noted above, together with a small community that can grant the passage of information across generations.

In the previous chapter we have also explained that, to educate an infant brain, rewards and punishments are needed: this fact indicates that organization can happen only through those two inputs and, finally, the unorganized human cortex is changed in an intelligent one thanks to *discipline* and *initiative*, which Turing considers the two main aspects of a process that has to be studied as it happens in humans to "copy it in machines" (Turing 1969, p. 21).

2.2.2 Educating Physical Entities

It is natural for Turing to employ the idea of education in the case of humans to construct the new concept of computation: also physical machines can be *educated* to produce certain kinds of modification. In this last case, education coincides with programming, which is an imitation of the human case: we can modify machines thanks to programming to the aim of reaching some good and interesting responses. The concept of (Universal) Logical Computing Machine (LCM) emerges in this epistemological atmosphere and, on the basis of this abstract theoretical tool, also the one of the (Universal) Practical Computing Machines (PCM). These are digital computers (as physical entities) that can be built and "educated", able to imitate the behavior of a human computer very well. I have already contended that, to show

Turing's externalization of both the abstract (the logical machine) and the concrete (the practical machine) entities, the expression *mimetic minds* is very appropriate (Magnani 2006, 2009): these machines are indeed capable of imitating the mind in a universal manner and are objectified in the environment, as clearly expressible formal intellectual structures and practical machines. They are universal because we do not need various machines that do different performances: thanks to suitable programming activities, we can have various performances done. The analogy in terms of education is completed saying that universal machines are subjected to *paper* interference, when the introduction of new information in the machine modifies its behavior, but also (and here the analogy is no longer active) to *screwdriver* interference, when some parts of the machine are eliminated and replaced by others, giving rise to completely new machines (by the way, in organic agents, this effect of changing the structure is caused by the evolution).[4]

2.3 Building Computational *Mimetic Bodies* Through Morphology-Based Enhancing

It is in the framework I have described in the previous section that we can limpidly see—naturally extending Turing's perspective—that the recent emphasis on the simplification of cognitive and motor tasks generated in organic agents by morphological aspects implies—once exploited in robotics—the need not only of further computational mimesis "*à la* Turing" of the related performances—when possible—but also the construction of appropriate *mimetic bodies* able to render the accompanied and integrated computation simpler, according to a general appeal to the *simplexity* of animal embodied cognition, which stresses possible complementary relationships between complexity and simplicity (Kluger 2008; Berthoz and Petit 2014). To comprehend this process, it is first of all necessary to illustrate various aspects of the so-called physical computation, in which we see exactly at play a further detailed description of that "organization of unorganized entities" introduced by Turing.

2.3.1 A Computer is a Physical System

In the first chapter of this book I have illustrated some aspects of the relationships between the concepts of information, cognition, and computation, trying to ask the question "Is cognition computation?" We can give a positive answer, it is, but it is not only computation, so that we cannot identify cognition with computation. Of course, information processing and computation are involved in cognition, and plenty of research has been done to clarify the various roles and types of computation

[4] More details on Turing's ideas summarized in this section and in this previous chapter are illustrated in Magnani (2009), Chap. 3.

and information processing implicated in cognition, even if these kinds of research, at least from an eco-cognitive perspective, are damned to become old-fashioned. I prefer to affirm, from a perspective informed by a kind of naturalistic epistemology, that, given the fact the concept of computation changes and is subject to meaning variations, the other two concepts, already individuated in themselves by precarious definitions, being associated with the developments of knowledge, technologies, and cultural frameworks, also vary.

In Sect. 1.1.3 of the previous chapter I have quoted recent studies which aim at elucidating the concept of digital computation, by illustrating how digital computation is implemented in *physical systems*: these studies do not end up in pancomputationalism,[5] that is, the view that every physical system is a digital computing one and can be described in computational terms. These studies mainly aim at philosophical/ontological and definitory results. As already said, my concern here is more humble: my eco-cognitive computationalism does not aim at furnishing an ultimate and static definition of the concepts of information, cognition, and computation. Instead, by valuing their historical and dynamical aspects, I will suggest an epistemological perspective that depicts how we can understand the change and extension of their meanings thanks to the example (and the description of the "emergence") of what I call the *new* computational "domestication" of physical entities, thanks to morphological computing. To this aim, I will adopt the perspective on physical computation offered by Horsman et al. (2014), p. 14, that is particularly appropriate to favor my concerns.

1. A *computer* is a physical system with actual constituent parts and its own internal interactions that take it from one physical state to another.
2. Hence, I agree with Horsman et al. who contend that physical computing is "the use of a physical system to predict the outcome of an abstract evolution" (Horsman et al. 2014, p. 14). Indeed, it is interesting to note that in computations we do not have to deal with a physical system that needs to be described (this is, for example, the case of physics), but, on the contrary, with an abstract object that we want to evolve thanks to the physical system itself. Once we have realized this exploitation of a physical entity, the physical evolution performs a computation, so interpreted by a human or another artificial agent.
3. A computer is usually a *technology* built by using our scientific theories to precision-engineer physical systems to desired specifications (I will illustrate more details on this issue in Sect. 2.3.3).
4. A physical system is computing when a special relationship of abstract mathematical/logical entities to the physical ones at stake is enforced.
5. Beyond conventional and unconventional cases, the notion of *computation*, and its related system property, *information*, has been imported into other fields in an attempt to describe and "explain" such diverse processes as photosynthesis and the conscious mind, and a strand of modern cross-disciplines have given us the claims that "everything is information" or "the universe is a quantum computer" or

[5] Cfr. also below, Sect. 2.4.

"everything computes" (Horsman et al. 2014, p. 2). Obviously, many researchers plausibly contend that defining the universe and everything in it as a computer, the notion of *physical computation* becomes *empty*. I will come back to this issue in Sect. 2.4.[6]

2.3.2 Physical Computing Versus Physics

As I have already anticipated, following the approach proposed by Horsman et al. (2014), physical computing can be usefully considered the inversion of mathematical science: *a physical system is used to predict the outcome of an abstract dynamics*, on the contrary, in physics, *an abstract model is used to predict physical dynamics*. Indeed, physics works by representing physical systems abstractly, using abstract theories to predict the result of physical evolution. Let us repeat: if physical computing is the inversion of mathematical science, and uses *a physical system to predict the outcome of an abstract dynamics*, without this predictive element, a physical system is not a computer, in the same way that a group of mathematical equations is a bad physical model if it does not have predictive power. Hence, in physical computation, we want to take an abstract entity, a computation, and represent it physically. It is a process of *representation* that grants a possible "comparison between physical processes and mathematical described computations" (Horsman et al. 2014, p. 2). On the contrary, in physics, for example, an electron is represented as a wave function, constituting a map between physical and abstract spaces.

In summary, in a computation, "the initial impetus is not a physical system that needs to be described, but rather an abstract object that we wish to evolve. An abstract problem is the reason why a physical computer is used" (cit., p. 10). We therefore start immediately with the problem of a reversed representation.

2.3.3 Computational Physical Systems Usually Are Technological Devices

The construction of what Turing called Universal "Practical" Computing Machines is not at all unrelated to the development of technologies able—in the case we are discussing—to "domesticate" physical artifacts to behave in a certain way: after all, with the aim of having at our disposal physical computations, we need appro-

[6] As an example of an alternative unconventional view of computation we can indicate the so-called Actor Model (Hewitt 2012): computation is conceived as distributed in space where computational machines communicate asynchronously and the whole computation is not in any well-defined state. In response to a message that it receives, an Actor can perform local decisions, generate more Actors, send new messages, and choose how to respond to the next message received. Turing's Model is a particular case of the Actor Model. In this perspective, the Internet performs unconventional computations and seems controlled by super-recursive algorithms (Dodig-Crnkovic 2011).

priate physical entities able to perform the task, and this is a fruit of technological advances. We have already observed in the previous chapter that is surely astonishing that humans were able to transport the Turing's abstract machine, as a Discrete-State Machine (DSM) (Longo 2009) to physical machines, creating artificial technological processes that are able to evolve by discrete states, thanks to the role played by silicon electronic devices. We also said that since Turing computer science teaches us that physical artefactual entities are built to process the organization of information and knowledge into little boxes, bits, and pixels, which present discrete accuracy, with no fuzziness and no contingency, contrarily to what is occurring in chaotic deterministic systems, in which a fluctuation/variation below the interval of measurement generates manifest different evolutions for a system. Turing notes in the 1950s that this consequence is theoretically evitable in his Discrete State Machine (DSM): The Turing machine is an abstract (logical, he said) machine, endowed with a Laplacian behavior, as a DSM the machine processes values that are inevitably discrete. The concept of a program and its mathematical characters that are related to its implementation are deterministic exactly in Laplace's sense: the use of rules (or in the case of Laplacian mechanics, of equations) entails predictability.

When a scientific theory in natural sciences is accepted (that is, when it is appropriately confirmed and relatively exempt of intolerable anomalies or falsifications), explanatorily powerful, and usually endowed with predictive capabilities, it is the prototype of the success of progressive scientific research program, as the philosopher of science Imre Lakatos clearly described (Lakatos 1970). We also perfectly know that these good scientific theories are at the basis of engineering: they are not only tools for discovering new physical systems (and to make predictions regarding their behavior), but also for constructing new artefactual physical systems. Scientific theories are used to engineer physical systems that humans can further "domesticate" to obtain technologies endowed with desired specifications and expected behaviors.

In the case of technologies, we do not have to deal with a theory that models a physical system but something reversed: it is necessary to start from a good physical theory $\mathcal{T}$, able to provide an abstract specification of the physical system that we aim at building that can be called $m_{p'}$. The aim of technology is to construct the corresponding physical system, $\mathbf{p}'$, effectively reversing the modeling relation: "The process of technology to produce this reversal consists of finding a physical system $\mathbf{p}$, the theory $\mathcal{T}$ and a specific set of evolutions $\mathbf{H}$ that will perform the evolution $\mathbf{p} \rightarrow \mathbf{p'}$ such that, when $\mathbf{p}'$ is represented using $\mathcal{R}_{\mathcal{T}}$, it becomes the desired $m_{p'}$.[7] The physical system $\mathbf{p}$ is thus engineered using the process $\mathbf{H}$ to produce the desired physical system $\mathbf{p}'$. An example would be taking a set of steel girders and building a bridge out of them" (Horsman et al. 2014, p. 9). Of course, finding the ways for reaching the final technological artifact requires a project that is the fruit of ingenuity and skill on the part of the various actors at play.

[7] Which is an abstract specification of the physical system that we wish to construct, within the representation of the theory at stake, to the aim of de facto reversing the modeling relation.

New Computational Substrates

We know that present standard substrates for computation are highly engineered silicon entities whose behaviors can be explained by very established and reliable physical theories. Now, we are dealing with the problem of finding new physical substrates able to carry computation, the first requirement is the availability of a reliable physical theory of them, able to guarantee good expected functions and their regularity. Because physical computing is the inversion of mathematical science, as I have indicated in the previous subsection, the physical system is used to *predict* the result of an abstract dynamics (rather than an abstract model predicting physical dynamics). Consequently, knowing the predictive capacities of the physical system is fundamental. Horsman et al. usefully describe an "unfortunate" method for attributing computational capacities to a non-standard system:

> A novel computing substrate is proposed (a stone, a soap bubble, a large interacting condensed-matter system, etc.) The physical substrate is "set going" and an evolution occurs. At a certain point, the end of the process is declared and measurements taken. The initial and final states of the system are compared, and then a computation and a representation picked such that if the initial state and final states are represented in such a way, then such a computation would abstractly connect them. The system is then declared to have performed such a computation: the stone has evaluated the gravitational constant, the soap bubble has solved a complex optimization problems, and so on (Horsman et al. 2014, p. 20).

In summary, the computational description of a physical evolution does not have to be enforced "only" post-hoc: "the challenge then for non-standard computation is to demonstrate that the theory of the device, and the representation of data within it, is known and stable enough to use the physical device to predict the desired abstract computation" (cit., p. 21).

2.3.4 Encoding, Decoding, Computational Entities, Unconventional Substrates

Four basic items are indispensable to illustrate the role of coding and decoding when exploiting physical computational entities and the status of their relationship with the presence of a "computational agent".

1. To start a computation, it is first of all necessary to initiate a process of *encoding*—which is an act of *representation*—abstract data and *actual embedding* the related problem in a physical system (indeed, we are assured that our artefactual physical system works because—as illustrated in the previous subsection—we have a good physical theory that can predict how encoded data will work). We have to add that encoding and actual embedding constitute the act of putting the problem into a form—thanks to some appropriate cognitive representation, for example mathematical—that can be carried by an abstract algorithm, which in turn through refinement and composition will be encoded into the machine

(the artefactual physical entity that starts to play the role of a computer), which will be able to manage it. It is through this process that the abstract algorithm becomes a concrete algorithm. In the meantime, abstract algorithms are transformed into equivalent concrete algorithms, exactly as Turing already described, and when the artefactual physical system is "domesticated" as a computer, and we can reliably expect to find the computable results of abstract evolutions, we finally face what Turing called universal practical machine (cf. Sect. 2.2). I have already observed that the act of encoding is subjective and related to a particular human or artefactual agent (indeed, the meaningfulness of encoding and decoding information in physical entities can be, for example, performed by non-biological but computational entities, such as an AI program).

2. Thanks to the physical evolution of the system—as I have said, these systems are usually artefactual highly engineered silicon devices with an extremely well developed physical theory in which we have a great deal of confidence—we arrive at the final physical state (which is *decoded* as another abstract state). Consequently, this final abstract state did not evolve abstractly, but concretely, physically. Without the encode and decode steps, there is no computation; there is simply an "ignorant"—so to speak—physical system undergoing evolution: "going about its business", *only potentially computational* (Horsman et al. 2014, p. 5). We can usefully add "This is how we can escape from falling into the trap of 'everything is information' or 'the universe is a computer': a system may potentially be a computer, but without an encode and a decode step it is just a physical system" (cit., p. 15). We have to remember that non standard physical artefactual entities that are used as computing devices and substrates—with the exception of quantum computers—have a theory much less developed and, consequently, the reliability is low.
3. Coding and decoding imply the need of a so-called *computational entity* (it can be organic or not: human, animal, artefactual) that is able to represent—and at the same time to delegate and to recognize *meaning* to—that specific physical system as that specific abstract object.
4. Can living organisms of all sorts be considered physical entities that potentially perform information processing, and that can potentially be exploited to perform their computations for us? As I will describe in the following subsection, the physical RC—Reservoir Computing—approach implies that we can exploit the dynamical properties of various kinds of body parts to carry out relevant computational tasks.

2.3.5 Morphology-Based Enhancement of Mimetic Bodies

Cognitive science has clearly illustrated that intelligent abilities in performing behaviors of natural agents is often explained with reference to their bodily structure (morphology). This morphology can be used for the engineering of intelligent abilities in artificial agents, such as robots. Literature has individuated three main roles played by morphology:[8]

1. morphology that facilitates control, where no control system, no motors and no sensors are present;
2. morphology that facilitates perception;
3. morphological computation in proper sense, such as "reservoir" computing, in which embodiment and computation are strictly connected.

In the first case, an example is the passive dynamic walker (in which the behavior is occurring by purely mechanical interaction); in the second case, the classical case is illustrated by the Gecko feet, which can in turn be considered as active extensions of the passive walker, thanks to actuators and sensors (the specific ability of the Gecko is the result of its morphology interacting with a particular environment—not primarily that of higher-level central control); also the case of the eye of the fly, which not only concerns movement, but also cognitive abilities related to perception, is another good example; the third case refers to the so-called Reservoir Computing, abstractly proposed by the neural network community, and optimally extended to the role of physical devices that can operate as a reservoir (Physical RC (Reservoir Computing)), in which the body is effectively domesticated for computation. Reservoir computing is not only limited to morphological computations, but, as I have just said, originates from research on neural networks, and contemplates a collection of neurons, as Müller and Hoffmann summarize,

> [...] with nonlinear activation functions and with recurrent connections that have a random but bounded strength; this is referred to as a dynamic reservoir. These neurons are randomly connected to input streams, and the dynamics of the input is then spread around and transformed in the reservoir, where it resonates (or "echoes"—hence the term "echo-state networks") for some time. It turns out that tapping into the reservoir with simple output connections is often sufficient to obtain complex mappings of input stream to output stream that can approximate the input–output behavior of highly complex nonlinear dynamical systems. During training, the weights from the input streams and between the reservoir neurons are left intact; only the output weights—from the reservoir to the output layer-are modified by a learning algorithm (e.g., linear regression). The complexity of the training task has been greatly reduced (as opposed to training all the connections) by exploiting the reservoir to perform a spatiotemporal transformation of the input stream (the temporal aspect of the input sequence has basically been unfolded by the reservoir and can be retrieved directly at any instant). Furthermore, if feedback loops from the output back to the reservoir are introduced and subject to training, the network can be trained to generate desired output streams autonomously (Müller and Hoffmann 2017, p. 6).

The amazing novelty is that some physical entities can work as a reservoir as well, giving origin to the so-called Physical RC, which can also be the body of an agent: indeed, biological bodies interacting with their environments can present the needed properties—high dimensionality, nonlinearity and fading memory.[9]

---

[8] See, for example, the rich treatment provided by Müller and Hoffmann (2017).

[9] More updated details can be found in Müller and Hoffmann (2017), p. 3 and Hauser et al. (2011, 2014).

2.3.6 The Birth of Mimetic Bodies: Enhancing Ignorant Bodies Through Distributed Computation

To further examine the ways of domesticating—beyond the usual world-wide widespread use of silicon devices in digital computing—non-stardard entities for computation, we need to describe some cognitive and epistemological aspects of the so-called RC (Reservoir Computing). In this case, a classical "digital" computational controller tool (also a robot for example)—that already needs its own "conventional" physical embodiment—is enhanced by the fact that some cognitive "virtues" are outsourced to the morphology of physical entities with the aim of getting the good fruits of an "analog" computation. Where should we draw the line between morphological computation and pure digital computation? Hauser et al. (2014, p. 227) furnish the explanation quoted below of the mechanisms of the RC (Reservoir Computing). Various real physical systems—certainly not explicitly built for computation—can potentially serve as reservoirs: examples of actual implemented reservoirs such as the well-known case of a soft silicone-based octopus arm, employed to emulate desired nonlinear dynamical systems and adopted to carry out computations and to implement a feedback controller, are illustrated in Hauser et al. (2011, 2014), Nakajima et al. (2015).

> At the core of RC lays the so-called *reservoir*, a randomly initialized high-dimensional, nonlinear dynamical system, which maps the typically low-dimensional input (stream) onto its high-dimensional state space in a nonlinear fashion. In that sense the reservoir takes the role of a kernel (in the machine learning sense, i.e., the nonlinear projection of a low dimensional input into a high-dimensional space [...]). In addition, the reservoir, being a dynamical system, has the inherent property to integrate input information over time, which is obviously beneficial for any computation that needs information on the history of its input values. It is important to note that the reservoir is not altered during the learning process. Although it is randomly initialized, its dynamic parameters are fixed afterwards. In order to learn to emulate a desired input output behavior (to be more precise, a desired mapping from input streams to output streams), one has to add a linear output layer, which simply calculates a linearly weighted sum of the signals of the high-dimensional state space of the reservoir. These output weights are the only parameters that are adapted during the learning process (Hauser et al. 2014, p. 227).

Implementation of the so-called Physical RC involves real physical bodies that are used as reservoirs and as computational sources. Obviously, both the entities involved in classical highly engineered silicon devices and in Physical RC exploitation of bodies do not "know" that they are part of a computational device. They simply conform to the laws of physics and they respond spontaneously. In the case of Physical RC, we have to note "[...] that this also implies that the body does not over—nor under—compensate, since it is a stable physical system. The proposed setup simply adds some linear readouts to the body to complete the computation. The body would react exactly the same, if there were no readouts at all. If this output is used, e.g., in a feedback loop as a control signal for the robot, the behavior of the robot of course should be different if we close the loop by adding the readout" (Hauser et al. 2014, p. 230). It is important to note that, in the case of morphological computation, a physical computer does not need to be intelligently conceived: it can be naturally (or even computationally) evolved. This means that living organisms or parts of organisms (and their artefactual copies) can potentially execute information processing, and can potentially be exploited to execute their computations for us.[10]

Hence, the body is so to speak "ignorant" (Hauser et al. 2014, p. 231), it behaves the same whether we add readouts or not and can be used in various computations on the same input at the same time adding the needed readout. Defining the body "ignorant" of the computations that are carried through it can sound awkward to some ears. However, determining the body as an ignorant term of the computation does not automatically imply that we can find a more knowledgeable part of it—nor it should point in that direction. On the contrary, considering the fact that no part of the system knows that is part of the system itself, I think that is significant to point out the "ignorant" feature of the body, as a fundamental part of the system, not with respect to the other "knowledgeable" parts, but in order to indicate a particular state of ignorance that we can attribute to it, which is not just lack of the permanence (and contamination) of information, but a form of passivity to the computation that happens through it.

Moreover, the classical "physical" computational entities (hardware) as silicon devices or robots endowed with an abstract controller[11] are based on the discreteness of *digital* processing and, from the point of view of their macroscopic features, are stable, unperturbable, and robust with respect to the external influences different from the processes of encoding and decoding (see Sect. 2.3.4). On the contrary morphological computation, in the case of Physical RC, is occurring in the continuous reality (without digital complication) as a clear form of very fast *analog* computation, in which the exploitation of real entities at the same time involves obvious limitations and constraints. We also have to note that to weaken the lack of what we call cognitive/computational plasticity (and—so to speak—of "universality" with respect to the classical Turing machines) of the physical adopted reservoir a kind of *morphosis* (just to make a simple example, moving the body into a new posture) (Hauser et al. 2014, p. 235) is necessary: the morphology could be modified online to get multiple computational benefits, also taking advantage of the possible effects on the whole system caused by the environment.

In summary, the bodies that are, as I illustrated above, "ignorant", in the framework of morphological computation become *mimetic bodies*, that is, bodies that are able to mime various cognitive mediating tasks. I call them mimetic exactly in Turing sense: Turing says by "mimicking education, we should hope to modify the machine until it could be relied on to produce definite reactions to certain commands" (Turing 1969, p. 14). Similarly, mimicking the morphological features we can get computational results: a standard example is furnished by Nakajima et al. (2013, 2015), who

---

[10] On this issue, cf. also (Horsman et al. 2014, pp. 19–22).

[11] Hauser et al. also usefully note that "Another important point is that classical robot designs (built out of rigid body parts connected with high torque servos) don't have any deliberate computation in the physical layer, hence, any computation in the robot is carried out only in the abstract controller layer" (Hauser et al. 2014, p. 234).

have "strictly" mimicked part of the morphology of an octopus using a model of a soft robotic arm inspired by it. Seeing to the new developments of morphological computation, we can surely conclude that we are facing new interesting epistemological and technological aspects of what we can call *distributed computation*: we see computation more and more distributed in a wide variety of props, tools, bodies, and devices to the aim of getting new cognitive results. The promise of morphological computation principles in robot design can originate a new generation of robots with better adaptability and restricted number of required control parameters.

2.4 Pancomputationalism Naturalized

The problems of pancomputationalism is nicely summarized in the following passage by Horsman et al. (2014, p. 2):

> There is currently no accepted answer to this question, and an absence of a worked out formalism within which to determine whether a computation is happening physically gives rise to a great deal of confusion when discussing non-standard forms of computation. We can all agree that a laptop running a Matlab calculation and a server processing search engine queries are physical systems performing computation. However, when we move beyond standard and mass produced technology, the question becomes more difficult to answer. Is a protein performing a compaction computation as it folds? Does a photon (quantum) compute the shortest path through a leaf in photosynthesis? Is the human mind a computer? A dog catching a stick? A stone sitting on the floor? One answer is that they all are – that everything that physically exists is performing computation by virtue of its existence. Unfortunately, by thus defining the universe and everything in it as a computer, the notion of physical computation becomes empty. To state that every physical process is a computation is simply to redefine what is meant by a "physical process"—there is, then, no non-trivial content to the assertion. A statement such as "everything is computation" is either false, or it is trivial; either way, it is not useful in determining properties of physical systems in practice.

The assumption of pancomputationalism clearly states that everything is computational (Müller 2017). This view is often defended contending that, when we say that everything is computational, we are just describing something as computational as a way of "interpreting" it, and everything can be interpreted that way. Unfortunately, this defense leads pancomputationalism to conflate computational modeling with computational explanation. Other forms of pancomputationalism contend that the whole universe can be considered computational and seem to also claim that information or computation have a kind of priority with respect to physical materiality (Piccinini and Scarantino 2011, p. 5). Dodig-Crnkovic (2011, 2013, 2017) proposes a richer info-computational view of the universe (a synthesis of pancomputationalism—naturalist computationalism—with informational structural realism) and defends it by observing the central role of computing in nature (natural computing).

As I have observed in the previous chapter, taking advantage of an evolutionary perspective and of Thom's concepts of the catastrophe theory, when we see the case of an infection as a pregnance (mediated by a virus, that is a material/biological medium) that affects healthy subjects, who are the invested saliences that in turn can re-emit that same contagion as a pregnance into the natural niche (in which, in turn, other media such as air or blood are the transmitters), it seems weird to contend that information (or computation) is at play, in this case against paninformationalism and pancomputationalism. One count is to produce a "model" of that event from an informational point of view or by using a computational program, and another count is to produce a biological knowledge of it.[12] However, I think that the positions that defend paninformationalism and pancomputationalism are significative for two reasons. The first reason is related to the consideration of natural human evolution, which presents that widespread activity of semiosis (including all kinds of signs, not only the propositional ones), that has been created by humans since the times of our primitive ancestors. In this way, humans have built those voluminous *cognitive niches*, hugely endowed with informational (and more recently, computational) processes (Odling-Smee et al. 2003; Laland and Sterelny 2006; Laland and Brown 2006), which certainly favor an inclination to envisage some ontological status to information and computation. In addition, I contend that thinking in terms of "distributed computation" (an expression I have introduced above in Sect. 2.3.6) helps us see pancomputationalism in a more naturalized way, avoiding ontological or metaphysical considerations.

The second reason is epistemological: in the literature, there is even an all-encompassing notion of information, a kind of paninformationalism, in which physical (or biological) information is extrapolated to every state of a physical (or biological) system that is delineated as an information-carrying state (Wolfram 2002; Lloyd 2006), which cannot be under-evaluated. Indeed, this perspective has favored many excellent results that physicists (for example) (Chiribella et al. 2012; Goyal 2011) (and logicians) have reached, for example providing mathematical frameworks for seeing quantum theory in the perspective of the principles of information processing. My only warning does not concern these results, but the possible abuse of the notions of information (and of computation) in physics and biology, a problem extensively illustrated in detail in Longo (2002, 2009), as I will describe in the following section.

2.5 Using Physical Computing to Model Physical and Biological Systems

As I have indicated above (Sect. 2.2), Turing argumentations are coherent with the illustration of physical computation as the realization in terms of a physical evolution of an abstract computation I described in Sect. 2.3. To quickly recall Turing's seminal ideas, we can say that he contends that a big cortex constitutes an evolutionary advantage only if it is fertilized by a great quantity of meaningful information and knowledge carried by external supports, props, and tools—interacting with it—that

[12] In the following section of this chapter, I will describe the dangers that can emerge by thinking that mimetic computational modeling "is" immediately, ipso facto, scientific knowledge.

only an already "evolved" collective of humans can have. In summary, *information*, but also *cognitive* contents carried thanks to language, signs, icons, etc. have to be more and more available to promote the useful exploitation of a big cortex. Paleoanthropologists such as Mithen (1996, 1999) would also add that storing in the external environment signs and drawings, and manipulating external entities transforming them in artifacts, is the main process that characterizes not only the birth of the so-called material culture, but also the trigger of a fundamental process of *coevolution* between brains and culture—on this issue cf. also the recent rich and informed book written by Laland (2017).

Making an analogy to this evolutionary process, Turing contends that transmuting an external physical artefactual entity (such as an electronic physical object) in a cognitive and computational one, "paper interference" is needed, when the introduction of new information in the machine modifies its behavior. The Turing lucubrations about the "unorganized" brains, described as kinds of blank slates that are socially fulfilled through language are really modern because, even if very abstract and characterized by a dominant heuristic role, show that phylogenetic mechanisms acting in human cognition are crucial and worth being taken into account, an attitude more powerful with respect to the one of traditional philosophical western schools.

As already said in the previous chapter, it is important to note that, in this view offered by Turing, the digital machine (a discrete state machine) is certainly an alphabetic machine: its conditions of possibility resort to human evolution towards alphabetic natural language. Longo (2009) contends that this fact is at the origin of that tremendous "discretization of knowledge" that Turing's achievements have created. The "continuous" natural language is indeed transmuted by the alphabet in something separated into small atoms, which forge letters. These atoms do not present any kind of meaning that instead comes out thanks to their syntactical aggregation made by skilled human agents able to sensibly combine them. This discreteness, typical of digital machines, is the fundamental aspect that motivates their *imitation* power—they are mimetic machines, *mimetic minds*, as I said—and Turing himself contrasted this simple imitation power to the much stronger epistemological power of the *modeling* capacity of mathematics, when he was thinking about the science of morphogenesis.

At this point, the problem of imitation leads us to consider a further aspect of physical computation: when computers are further used to model physical (or biological systems), are we still dealing with imitation or with a kind of reliable production of knowledge that could occasionally be called "scientific"? Let us see more details concerning this problem. We have indeed said in the previous section that physical computation is a transposition of physical evolution for abstract computation, that is, we can say, following Turing, that we "educate" a physical system to perform a computation. Once this task is performed, we can in turn submit a physical (or biological) dynamic of a system to a computational modeling, that is, we can use computers to simulate the behavior of a physical or a biological system. In the meantime, it is important to note that in this case the computational simulator and the physical or biological system simulated interact at the abstract level and what is simulated is a *model* of the physical or biological system, not the system itself

(Horsman et al. 2014, p. 17). We know that computational modeling of physical or biological processes can be extremely useful as a heuristic tool in actual scientific research, also at the creative level (just to make a simple example, to simulate the behavior—through modeling (Magnani and Bertolotti 2017)—of a physical system during an experiment), but we have to note that the study of a physical or a biological evolution can not always take advantage of a computational representation. Hence, what is the epistemological status of computational simulation?

The dichotomy between discreteness and continuity involves a reflection upon the other related dichotomy between imitation (as an effect of the computational representation) and intelligibility (as the fruit of scientific knowledge).

We have said few lines above that the computational turn has generated a huge "discretization of knowledge". This discretization, in the case of the mental representations of "concepts", generated great consequences. We can guess that an isomorphism is established between the mental processes (where of course "phonemes" play a dominant role), which ensure the stability of a concept, and the physical and material processes that ensure the stability of the actual object represented by the concept. A discretization of knowledge that—long before the computational turn—did not have a marginal role in the recent history of human cognition. The internalization of phonemes (Coolidge and Wynn 2005) has been related to the development of the "enhanced working memory" (EWM) appeared about 30,000 years in hominids and seemed to coevolve with the emergence of a *phonological storage capacity* along with consequent language and other modern reasoning abilities such as planning, problem-solving/algorithm manipulation, analogy, modeling, holding inner representations, tool-use, and tool-making, but also cross-modal thinking needed in social tasks. In particular, an increase in phonological storage could also have aided cross-modal thinking (and so hypothetical cognition) and the social tasks caused by the need for the preservation and defense of collective coalitions: "[…] enhanced phonological storage may have freed language from the laconic and its confinement to present tense and simple imperatives to rapidly-spoken speech and the use of future tense—the linking of past, present, and future, and the use of the subjunctive […]. Although real enemy's actions might be anticipated, imaginary enemies could be envisioned and other intangible terrors could be given life. Great anxieties could arise with novel vistas (e.g., the meaning of life, thoughts of death, life after death, etc.)" (cit., p. 22).

Notwithstanding the triumph of discretization, in western written natural languages but also in philosophical and logical knowledge, from Democritus to Descartes and from the modern XIX century axiomatics to the computational turn, we are still facing the conundrum represented by the fact that, however, this simple reality of small components is actually very complex, as recent scientific research into the dynamical systems theory, quantum physics, and biology demonstrates. For example, it is difficult to study the cell only by referring to its constituents, and also its "wholeness" is fundamental. The suspect is that, when we use computational devices, which are discrete machines, to produce knowledge about physical and biological systems, some serious expressive limits arise: it is unlikely that these machines can play the role of instruments for directly building scientific intelligibility regarding complex objects/systems such as physical and biological entities (and also human cognitive systems themselves). We also have to finally note that, in the last few decades, as I have already observed in the previous chapter, the notion of digital computation has been very important to the aim of modeling neural activity and other relevant features of human cognition, but some neuroscientists (for example Port and van Gelder 1995) convincingly maintain that neural networks do not perform digital computation at all: in this last case, cognition is obviously seen as being rendered possible by neural networks but is not computation and the approach we need to use to study cognition has to be related to the theory of dynamical systems.

The deep difference between the idea of scientific knowledge as simple *imitation*, as developed by digital machines and their "modeling capacity", and the idea of scientific knowledge as generation of *rational intelligibility* is a core problem of the dynamical approach (Port and van Gelder 1995). As I already said above, Turing himself emphasized the difference between the simple *imitation* capacity of machines and the *modeling* power of mathematics: the double pendulum, which is a perfect deterministic machine, only expressed by two equations, is sensible to minor variations, below the threshold of observability, it is a typical chaotic deterministic system, and it is extremely difficult to represent it by a mimetic machine (Longo 2009). Longo also notes that this is a system sensitive to initial conditions. It can instead be described using the mathematics of deterministic chaos, in which determination does not involve predictability: in this system, a process does not follow the same trajectory even if we reiterate it with the same initial conditions, within the limits of physical measures (*ibid.*) Furthermore, we can also usefully observe that, in the case of Turing machines that simulate such systems, when we restart after having processed analog simulations, using the same initial data, the same already seen trajectory is performed and, on the contrary, as said, the "real" pendulum behavior is completely different, when we restart the pendulum never performs the same trajectory.

Only "mathematical" models can explain the structure of physical causality of the system at play, and digital simulation only *resembles* or *imitates* causality, always realizing Laplacian and predictable processes. The explanation of this conundrum resorts to the fact that, at the roots of digital data, there is a *discrete* topology, but physical measurement is always an interval that is optimally represented by continuous mathematics. Even if sometimes the use of mathematical equations in physics does not have predictive power, this provides a knowledge characterized by a rich *qualitative* epistemological value, able to make intelligible the physical causality of the system (Longo 2012). Finally, we have to note that, with respect to physics, in the case of biological organisms, the gap between simulation and intelligibility is even worse because the variability is dominant. Longo (2017) contends that, "in biology, in particular, the introduction of information as a new observable on discrete data types has been promoting a dramatic reorganization of the tools for knowledge" and that some consequences of this effect have been induced in life sciences, with particular emphasis on research on cancer.

In summary, digital simulation—even if epistemological useful at the level of intermediate modeling during the processes of scientific research and discovery, as

I have already indicated above in this section—tends to produce epistemologically distorting results due to its simply mimetic quality.

Summary

In this chapter, taking advantage of an approach I called "eco-cognitive computationalism", I have first of all considered again Turing's original intellectual perspective that furnished the conceptual framework able to show how, thanks to an imitation of the evolutionary emergence in humans of information, meaning, and of the first rudimentary forms of cognition, the subsequent invention of the Universal Practical Computing Machine is achieved, that computer that, from the perspective offered by Turing, I call "mimetic mind". In the second part of the chapter, I have exploited this framework to illustrate the recent results of morphological computation in which the emphasis on the simplification of cognitive and motor tasks has rendered possible the construction of appropriate "mimetic bodies" able to render accompanied computations simpler, according to a general appeal to the "simplexity" of animal embodied cognition. I have stressed that, in the case of morphological computation, we can surely conclude that we are facing a new activity of what we can call "distributed computation": the promise of morphological computation principles in robot design can originate a new generation of robots with better adaptability and restricted number of required control parameters. Finally, I have also dedicated a short discussion to the concepts of paninformationalism and pancomputationalism, showing that the framework of distributed computation helps us see them in a more naturalized and prudent perspective, avoiding ontological or metaphysical considerations. The last section of the chapter is devoted to illustrating the related problems regarding the epistemological limitations of computation modeling when used to simulate the behavior of a physical or a biological system (Horsman et al. 2014, p. 17).

References

Berthoz, A., Petit, J.-L. (eds.): Complexité-simplexité. Collège de France, Paris (2014)

Chiribella, G., D'Ariano, G.M., Perinotti, P.: Quantum theory, namely the pure and reversible theory of information. Entropy **14**, 1877–1893 (2012)

Coolidge, F.L., Wynn, T.: Working memory, its executive functions, and the emergence of modern thinking. Cambridge Archealog. J. **5**(1), 5–26 (2005)

Dodig-Crnkovic, G.: Significance of models of computation, from Turing model to natural computation. Minds Mach. **21**(2), 301–322 (2011)

Dodig-Crnkovic, G.: The info-computational nature of morphological computing. In: Müller, V.C. (ed.) Philosophy and Theory of Artificial Intelligence, pp. 59–68. Springer, Berlin, Heidelberg (2013)

Dodig-Crnkovic, G.: Computational dynamics of natural information morphology, discretely/continuous. Philosophies **2**(4) (2017)

Goyal, P.: Information physics—towards a new conception of physical reality. Information **3**, 567–594 (2011)

Hauser, H., Füchslin, R.M., Nakajima, K.: Morphological computation—the physical body as a computational resource. In: Hauser, H., Füchslin, R.M., Pfeifer, R. (eds.) Morphological Computation: The Body as a Computational Resource, pp. 226–244. Self-published, Bristol (2014)
Hauser, H., Ijspeert, A.J., Füchslin, R.M., Pfeifer, R., Maass, W.: Towards a theoretical foundation for morphological computation with compliant bodies. Biol. Cybern. **105**(5), 355–370 (2011)
Hewitt, C.: What is computation? Actor model versus Turing's model. In: Zenil, H. (ed.) A Computable Universe. Understanding and Exploring Nature as Computation, pp. 159–185. World Scientific, Singapore (2012)
Horsman, C., Stepney, S., Wagner, R.C., Kendon, V.: When does a physical system compute? Proc. R. Soc. A, **470**, 1–25 (2014). 470 20140182; Published 9 July 2014. https://doi.org/10.1098/rspa.2014.0182
Hutchins, E.: Cognition in the Wild. The MIT Press, Cambridge, MA (1995)
Hutchins, E.: Cognitive ecology. Topics in Cognitive Science **2**(4), 705–715 (2010)
Kluger, J.: Simplexity. Why Simple Things Become Complex (and How Complex Things Can Be Made Simple). Hyperion Books, New York, NY (2008)
Lakatos, I.: Falsification and the methodology of scientific research programs. In: Lakatos, I., Musgrave, A. (eds.) Criticism and the Growth of Knowledge, pp. 365–395. The MIT Press, Cambridge, MA (1970)
Laland, K.N.: Darwin's Unfinished Symphony: How Culture Made the Human Mind. Princeton University Press, Princeton (2017)
Laland, K.N., Brown, G.R.: Niche construction, human behavior, and the adaptive-lag hypothesis. Evol. Anthropol. **15**, 95–104 (2006)
Laland, K.N., Sterelny, K.: Perspective: Seven reasons (not) to neglect niche construction. Evol. Int. J. Organic Evol. **60**(9), 4757–4779 (2006)
Lloyd, S.: Programming the Universe: A Quantum Computer Scientist Takes on the Cosmos. Knopf, New York, NY (2006)
Longo, G.: Laplace, Turing, et la géométrie impossible du "jeu de l'imitation": Aléas, determinisme et programmes dans le test de Turing. Intellectica **35**(2), 131–61 (2002)
Longo, G.: Critique of computational reason in the natural sciences. In: Gelenbe, E., Kahane, J.-P. (eds.) Fundamental Concepts in Computer Science. Imperial College Press/World Scientific, London (2009)
Longo, G.: Incomputability in physics and biology. Math. Struct. Comput. Sci. **22**(5), 880–900 (2012)
Longo, G.: The biological consequences of the computational world: Mathematical reflections on cancer biology (2017). arXiv.org, eprint arXiv:1701.08085
Magnani, L.: Mimetic minds. Meaning formation through epistemic mediators and external representations. In: Loula, A., Gudwin, R., Queiroz, J. (eds.) Artificial Cognition Systems, pp. 327–357. Idea Group Publishers, Hershey, PA (2006)
Magnani, L.: Abductive Cognition. The Epistemological and Eco-Cognitive Dimensions of Hypothetical Reasoning. Springer, Heidelberg/Berlin (2009)
Magnani, L.: Computationalism in a dynamic and distributed eco-cognitive perspective. In: Dodig-Crnkovic, G., Burgin, M. (eds.) Philosophy and Methodology of Information, vol. 1, pp. 365–288. World Scientific, Singapore (2018)
Magnani, L., Bertolotti, T. (eds.): Handbook of Model-Based Science. Springer, Switzerland (2017)
Mithen, S.: The Prehistory of the Mind. A Search for the Origins of Art, Religion, and Science. Thames and Hudson, London (1996)
Mithen, S.: Handaxes and ice age carvings: Hard evidence for the evolution of consciousness. In: Hameroff, A.R., Kaszniak, A.W., Chalmers D.J. (eds.) Toward a Science of Consciousness III. The Third Tucson Discussions and Debates, pp. 281–296. MIT Press, Cambridge (1999)
Müller, V.C.: Pancomputationalism: Theory or metaphor? In: Hagengruber, R., Riss, U. (eds.) Philosophy, Computing and Information Science, pp. 213–221. Pickering & Chattoo, London (2017)

Müller, V.C., Hoffmann, M.: What is morphological computation? On how the body contributes to cognition and control. Artif. Life **23**(1), 1–24 (2017)

Nakajima, K., Hauser, H., Kang, R., Guglielmino, E., Caldwell, D., Pfeifer, R.: A soft body as a reservoir: Case studies in a dynamic model of octopus-inspired soft robotic arm. Front. Comput. Neurosci. **7**, 91 (2013)

Nakajima, K., Hauser, H., Li, T., Pfeifer, R.: Information processing via physical soft body. Sci. Rep. **5** (2015)

Odling-Smee, F.J., Laland, K.N., Feldman, M.W.: Niche Construction. The Neglected Process in Evolution. Princeton University Press, Princeton, NJ (2003)

Piccinini, G., Scarantino, A.: Information processing, computation, and cognition. J. Biol. Phys. **37**(1), 1–38 (2011)

Port, R.F., van Gelder, T. (eds.): Mind as Motion. Explorations in the Dynamics of Cognition. The MIT Press, Cambridge, MA (1995)

Turing, A.M.: Intelligent machinery [1948]. In: Meltzer, B., Michie, D. (eds.) Machine Intelligence, vol. 5, pp. 3–23. Edinburgh University Press, Edinburgh (1969)

Wolfram, S.: A New Kind of Science. Wolfram Media, Champaign (2002)

Chapter 3
AlphaGo, Locked Strategies, and Eco-cognitive Openness

Locked and *unlocked strategies* are at the center of this chapter, as ways of shedding new light on the cognitive aspects of deep learning machines. The character and the role of these cognitive strategies, which are occurring both in humans and in computational machines, is indeed strictly related to the generation of cognitive outputs, which range from weak to strong levels of knowledge creativity. For a better understanding of the computational aspects I will also illustrate the abductive character of the concept of *anticipation* in human cognition linking it to the cognitive problem of spatiality and of the genesis of space in the description of the abductive role of the *Abschattungen* (adumbrations), as they are described in the framework of the philosophical tradition of phenomenology. Anticipations share various features with visual and manipulative abduction and prove to be a useful tool to favor the characterization of the two kinds of strategic reasoning I am introducing in this chapter, very important from the computational point of view. It will become clear that the differences between strong and weak levels of creativity lead to important consequences when we analyze computational AI programs, such as AlphaGo/AlphaZero, which aim at performing various kinds of abductive hypothetical reasoning. In these cases, the programs are characterized by *locked* abductive strategies: they deal with weak (even if sometimes amazing) kinds of hypothetical creative reasoning, because they are limited in what I call *eco-cognitive openness*, which instead qualifies human cognizers who are performing higher kinds of abductive creative reasoning, where cognitive strategies are instead *unlocked*.

L. Magnani, *Eco-Cognitive Computationalism*, Cognitive Systems Monographs 43,
https://doi.org/10.1007/978-3-030-81447-2_3

3.1 Are AlphaGo Cognitive Strategies Locked? An Abductive Framework

In 2015, Google DeepMind's program AlphaGo (able to perform the famous Go game) beat Fan Hui, the European Go champion and a 2 dan (out of 9 dan) professional, five times out of five with no handicap on a full-size 19 × 19 board. Later on, in March 2016, Google also challenged Lee Sedol, a 9 dan player who is considered the top world champion, to a five-game match. The DeepMind program defeated Lee in four of the five games. It was said that the program "invented" a new and unconventional move—never adopted by human beings—which was able to originate a new strategic framework, phenomenologically seen as simulating a proper "human" skillful ability, better than the ones of the more experienced humans. It is well-known that AlphaGo acquired the capacity to play the game by taking advantage of "seeing" data of thousands of games, perhaps also including those played by Lee Sedol, exploiting the so-called "reinforcement learning": the program in turn plays successively against itself to improve, enrich, and adjust once again further its own deep neural networks grounded on trial and error procedures. I have to add that the machine also implicitly actualizes what I call "cognitive strategies" to simplify the search space for the next best move: it is in this way that an almost infinite range of chances can be reduced to a more manageable quantity.

Let us adopt the meaning of the attribute "strategic" to refer—this seems particularly appropriate in the case of AI studies—to an expert mixture in reasoning of various heuristic cognitive devices: a strategy consists in a subsequent smart choice of the next state of a cognitive routine (for example, the nearest one, by respecting an appropriate distance measure) and a heuristic is one of the instruments a strategy can exploit to rapidly arrive to a wanted state. When we are referring to the non-computational case of game theory, the meaning of the word strategy more amply refers to the role of the agents in their relationships with other agents and to the various related contentious, interlaced, or collective cognitive acts. In the field of research of the so-called ecological thinking (or ecological rationality) (Gigerenzer and Selten 2012; Raab and Gigerenzer 2005; Gigerenzer and Brighton 2009), strategies acquire an even more extended meaning: they also regard thinking routines which take advantage of a great amount of information, knowledge, and elevated computational costs; on the contrary heuristics are used to execute more simple and efficient cognitive processes, even if less precise. Sometimes cognitive heuristics are more generically seen as coincident with cognitive strategies. In this chapter, it is useful to adopt the AI more widespread perspective: strategies consist in a wise successive choice of suitable heuristics.

I contend that it is in the framework of abductive cognition (Magnani 2009, 2017) we can appropriately and usefully analyze the concept of strategic reasoning to the aim of seeing the distinction between *locked* and *unlocked strategies*. Indeed, I will contend that in AlphaGo only locked strategies are at plays, because the elements of each scenario are always the same (what changes is merely the number of seeable stones and their dispositions in the board), in a finite and stable framework (no new rules, no new objects, no new boards, etc.), and this fact affects the type of creativity which is in general performed by deep learning machines.

In my studies on abduction, I have extendedly illustrated various kinds of human (but also computational) hypothetical cognition. I suggested the adoption of the dichotomous distinction between *selective* abduction (Magnani 2001)—for example, in diagnosis (in which abduction is basically described as an inferential process of "selecting" from a "repository" of pre-stored hypotheses)—and *creative* abduction (abduction that produces *new* hypotheses). Furthermore, I have shown that abduction is not only related to the propositional aspect, i.e. when processed using human language (oral and written), but can also be "model-based" and "manipulative". In the first case, we deal with an abduction that is basically performed thanks to internal cognitive acts that take advantage of models – also external – such as simulations, visualizations, diagrams, etc.; in the second case the external dimension is at play: in this case, an eco-cognitive perspective is fundamental because we have to refer to those cognitive actions (embodied, situated, and enacted) in which the role of external models (for example artifacts), props, and devices is mandatory, and in which the characters of the actions themselves are hidden and hard to be extracted. Action can give origin to otherwise unavailable data that makes the agent solve problems by initiating and processing an appropriate abductive process of production and/or selection of hypotheses. As I say, manipulative abduction is occurring when we are thinking "through" doing and not only, in a pragmatic sense, about doing (cf. (Chap. 1 Magnani 2009)). It is clear that, when we are dealing with games such as Go, manipulative abduction is also at play, given the fact the reasoning is considerably intertwined with the manipulation of the stones and various embodied aspects are involved, together with the visualization of the whole scenario, the adversary, etc.

3.1.1 Abduction and AI

The concept of abduction has been involved in AI at least since the beginnings of this young discipline. Already in 1988 Thagard (1988) described four types of abduction implemented in PI, a computational program devoted to performing some of the main cognitive abilities illustrated by philosophy of science: scientific discovery, explanation, evaluation, etc.[1] The program explicitly executes the so-called simple, existential, rule-forming, and analogical abduction. In this case the use of a computer simulation exhibited a first sophisticated new tool to increase knowledge about abduction itself, showing how this kind of non-deductive reasoning can be automatically rendered thanks to computational concrete artifacts.

[1] The AI research on these topics also favored the formation, in two philosophy departments, of the following facilities: the Computational Epistemology Laboratory (http://cogsci.uwaterloo.ca/) headed by P. Thagard at the University of Waterloo, Canada and the Computational Philosophy Laboratory (http://www-3.unipv.it/webphilos_lab/wordpress/), headed by myself at the University of Pavia, Italy, both devoted to research into computer science, cognitive science, and related areas of philosophy.

Early works on *machine scientific discovery*, such as the well-known Logic Theorist (Newell et al. 1957), DENDRAL in chemistry (Lindsay 1980), and AM in mathematics (Lenat 1982), demonstrated that *heuristic search* in combinatorial spaces represents an appropriate and general instrument for automating scientific discovery, and abduction was explicitly or implicitly categorized. In 1995, the AAAI Society organized a Spring Symposium on "Systematic Methods of Scientific Discovery" and in 1997 the Journal *Artificial Intelligence* promoted a special issue on "Machine Discovery" (91(2), 1997, cf. the classical Simon et al. 1997 and Okada and Simon 1997).[2] In 1990, the AAAI Society coordinated a Spring Symposium on "automated abduction", in which various computational programs able to execute various abductive tasks were analyzed.[3] In these seminal programs, abduction was mainly rendered at a symbolic/propositional level, using rules and heuristics, and logic programming prevailed.[4]

Already in 1992 Boden (1992) illustrated the distinction between classical programs able to *re-generate* historical cases of scientific discovery in physical science (Simon's five BACON systems and GLAUBER, Langley et al. (1987)), and systems able to build *new* discoveries (DENDRAL and AM, cited above). Other authors (for example, Langley and Klahr (1995), who wrote the program ACT-R) stressed the further distinction between computational systems that regard the processes of abductive hypothesis formation and of evaluation: PHINEAS (Falkenhainer 1990); AbE (O'Rorke et al. 1990); ECHO (Thagard 1989, 1992); TETRAD (Glymour et al. 1987), and the already quoted MECHEM. Other programs addressed the abductive nature of experimental procedures (DEED (Rajamoney 1993); DIDO (Scott and Markovitch 1993)), and, finally, we have to remember the ones that addressed both processes of hypothesis formation and evaluation, and of experiment (KEKADA (Kulkarni and Simon 1988); SDDS (Klahr and Dunbar 1988), LIVE (Shen 1993)). Some of the classical computational programs I have quoted above modeled what I call sentential creative abduction (BACON and GLAUBER in terms of heuristic search reproduced famous discoveries in physics), others performed model-based creative abduction (the already quoted AbE, PHINEAS and TETRAD, and CHARADE (Corruble 1997), GALATEA (Davies and Goel 2000), and PROTEUS (Davies et al. 2008), regarding causal and analogical abductive reasoning).

---

[2] Classical volumes where the reader can find the illustration of the most important research and of some historical machine discovery programs are Langley et al. (1987) and Shrager and Langley (1990). Cf. also Zytkow (1992) (Proceedings of MD-92 Workshop on "Machine Discovery"), and Colton (1999) (Proceedings of AISB'99).

[3] The proceedings are still available online https://www.researchgate.net/publication/235198935_Working_Notes_of_the_1990_Spring_Symposium_on_Automated_Abduction..

[4] A review of the classical AI approaches to abduction (mainly based on logic programming) is given by Paul (2000) and (Bylander et al. 1991; Reiter 1987; Kleer and Williams 1987; Reggia et al. 1983) (*set covering* approaches). Other classical programs regarding discovery in science are illustrated by Valdés-Pérez (1999): MECHEM (reaction mechanisms in chemistry (Zeigarnik et al. 1997)), ARROSMITH (intertwining between drugs or dietary aspects and diseases in medicine (Swanson and Smalheiser 1997)), GRAFFITI (generation of conjectures in graph theory and other mathematical areas (Fajtlowicz 1988)), MDP/KINSHIP (determination of classes within a classification in linguistics (Pericliev et al. 1998)).

Finally, as I already said, there are also programs able to represent the activity of the experiment (which is close to the processes of what I call "manipulative abduction"). Already in 2008 Clark himself (Clark 2008 p. 202) was convinced that artificial intelligence and robotics would have also involved in the near future various aspects related to embodiment, action, situatedness, and manipulation of external entities appropriately cognitively reshaped:

> The increasingly popular image of functional, computational, and information-processing approaches to mind as flesh-eating demons is thus subtly misplaced. For rather than necessarily ignoring the body, such approaches may instead help target larger organizational wholes in ways that help reveal where, why, how, and even how much [...] embodiment and environmental embedding really matter for the construction of mind and experience.

Other computational programs that demonstrated their efficaciousness in the execution of machine discovery abductive tasks derived from the studies on genetic algorithms and evolving neural networks (cf. for example (Pennock 1999, 2000)—in which creative abductive reasoning is rendered by exploiting some of the Darwinian mechanisms involved by evolutionary theories) and also from the so-called research on DNA computers (Boneh et al. 1996). Finally, we have to remember that other traditional AI programs have been built to reproduce abduction in mathematical and geometrical reasoning, such as ARCHIMEDES (Lenat 1994; Lindsay 1998, 2000b, a) and HR (Colton 1999; Pease et al. 2005).[5]

Traditionally, philosophy, epistemology, and logic have been the main disciplines interested in studying human reasoning. However, reasoning is by now also a major subject of investigation in AI and in the whole area of cognitive science. In this perspective, we can say that many fundamental cognitive processes (such as abduction, disregarded by the mainstream deductive logic tradition), when implemented in a computer, become AI programs: the abstract theories and the more concrete computational programs become two different ways of expressing the same thing. In this sense, theories of reasoning are about rules for reasoning and these are rules that teach us to do certain things in certain situations. Writing an AI program permits us to delineate these rules accurately. Computational programs that execute various kinds of hypothetical reasoning can be seen as prosthetic "abducers": just as microscopes are technologically created to widen the sensory endowments of humans, computational models of abductive cognition are created to extend human cognitive capacities.

What about the new perspectives on hypothetical abductive reasoning offered by deep learning programs such as AlphaGo? As I have anticipated, to clarify the cognitive character of this program, the examination of the kinds of strategies that are at play is in my opinion central.

[5] A rich survey of the intertwining between computation and scientific explanation and abductive discovery is illustrated in Thagard and Litt (2008).

3.1.2 Abduction and AlphaGo

It is now clear that studies on abduction are very useful when we have to analyze creative reasoning, and a simple *new* and unexpected move of a human being who is playing a Go game surely represents a kind of creative reasoning. In this chapter, it will be the keystone concept of *knowledge-enhancing abduction* and the related one of *eco-cognitive openness* that will favor a deep understanding of the logical and cognitive status of those kinds of cognitive strategies I will describe and that I call *locked* and *unlocked abductive strategies*. Locked and unlocked strategies are at the center of this chapter, as ways of shedding new light on the cognitive aspects of deep learning machines. The character and the role of these cognitive strategies, which are occurring both in humans and in computational machines, is indeed strictly related to the generation of cognitive outputs, which range from weak to strong level of knowledge creativity. I maintain that these differences lead to important consequences when we analyze computational AI programs, such as AlphaGo, which aim at performing various kinds of abductive hypothetical reasoning.

These programs are characterized by *locked* abductive strategies: they deal with weak (even if sometimes amazing) kinds of hypothetical creative reasoning, because they are limited in what I call eco-cognitive openness, which instead qualifies human cognizers who are performing higher kinds of abductive creative reasoning, where cognitive strategies are instead *unlocked*.

An objection to the adoption of the concept of abduction to shed more cognitive light on the strong impact of deep learning programs such as AlphaGo in contemporary AI regards the fact that they are based on hierarchical neural networks that operate on a subconceptual level: abduction has been instead fundamentally investigated thanks to symbolic formal models related to the tradition of logic.[6] I have indicated in the previous subsection that certainly the concept of abduction enters in the last part of the previous century traditional AI research thanks to the studies concerning automated scientific discovery (creative abduction) and medical diagnostic reasoning (selective abduction). The dominant representational tools were the symbolic ones such as classical logic programming, rule-based systems, probabilistic networks, etc. Can good abductive processes be modeled using representational tools and algorithms that operate on a subconceptual level? The answer is yes.

Bruza et al. (2006) insisted that it would be misguided to adopt a simple traditional, symbolic perspective of an abductive logical system by assuming a propositional knowledge representation and proof-theoretic approaches for driving it, because this perspective seems conceptually incomplete insofar as it ignores what is going "down below" (Gärdenfors 2000), which can be interpreted as the subconceptual level of cognition. They proposed semantic spaces as a computational approximation of Gärdenfors' conceptual space. Abductive hypotheses generated from semantic spaces do not have a proof-theoretic basis, but rather they are computations of associa-

[6] A simple neural network has been used to build the computational program ECHO (Explanatory Coherence) regarding that part of abduction that concerns the process of hypothesis evaluation (Thagard 1989, 1992).

tions by various means within the space: the passage from the sub-conceptual to the conceptual level usually involves a reduction of the number of dimensions that are represented. Not only from a cognitive perspective it seems that interesting hypotheses and dimensional reduction are inextricably related as information passes from the subconceptual to conceptual level. More recently, Bruza et al. (2012) interestingly analyzed concept combinations in human cognition by showing emergent associations still as the result of abductive reasoning within conceptual space, below the symbolic level of cognition, but in terms of a tensor-based approach which conceptualizes combinations as interacting quantum systems.

In sum, certainly the mainstream non-standard logical tradition which created models of abductive inferences was characterized by *symbols* but, from a more extended cognitive and philosophical perspective, also *multimodality* (that is cognition in terms of non-propositional models, icons, thought experiments, simulations, etc.),[7] and *implicit reasoning* appear to be important. Moreover, I have to remember that, from a wide cognitive and philosophical perspective, as I have illustrated in my own research (Magnani 2009, 2017), the term abduction refers to all kinds of cognitive activities that lead to hypotheses, in human and non-human animals. For example, humans often generate abductive hypotheses thanks to manipulative, embodied, and unconscious endowments, and higher mammals surely do not take advantage of symbolic syntactic language but instead other multiple cognitive capacities. Analogously to what is happening in the case of humans, that can perform abductions in various ways, there is not in AI a unique method able to favor the development of programs able to reproduce abductive cognition to hypotheses. Various knowledge representation formalisms and algorithms can be adopted to implement an appropriate computational program.[8]

At this point, we can go ahead and try to analyze the specific kind of abductive performance (the generations of "moves") that characterizes the deep learning AI program AlphaGo.

3.2 Natural, Artificial, and Computational Games

3.2.1 Locked and Unlocked Strategies in Natural and Artificial Frameworks

Go is a game played by human agents and AlphaGo is a computational deep learning program that can play the role of an automatic agent/player, so that competition with humans can become partially computationally determined. Go is already an "artificial" game, as it is invented by human beings and, consequently, takes advantage of

[7] A survey about the importance of models in abductive cognition is illustrated in Figueroa (2012).

[8] The need of a plurality of representations was already clear at the time of classical AI formalisms, when I was collaborating with AI researchers to implement a Knowledge-Based System (KBS) able to develop medical abductive reasoning (Ramoni et al. 1992).

abstract rules and artifacts, such as the board and other material objects. AlphaGo is artificial too, but a more complicated fruit of the technological creativity of few human beings.[9] However, we have to remember that also "natural cognitive games", so to speak, can be contemplated. For example, as I have already illustrated in the previous section when delineating manipulative abduction, a strategic human cognition not only refers to propositional aspects concerning acts performed through written and spoken language, but it is also active in a distributed cognition environment, in a kind of "game" in which models, artifacts, internal and external representations, sensations, and manipulations play a central function: imagine the pre-linguistic cognitive "natural game" between humans and their surroundings, in which "unlocked" strategies (see below) are at play, such as the phenomenological tradition has illustrated, exactly involving embodied and distributed systems, and visual, kinesthetic, and motor sensations (Magnani 2018).

What counts here is that in the natural games the strategies are *unlocked* because, even if local constraints are always at play in the interaction humans/environments, no predetermined background is established. On the contrary, what happens in the case of human-made "artificial games" such as Go, or in the case of their computational counterpart, such as AlphaGo? In these two last cases, the involved cognitive strategies are *locked*, as I will describe in the following paragraphs. However, before reaching the problem of computational strategies it is useful to analyze in the next two subsections the role of unlocked strategies in the interaction humans/environments.

3.2.2 Adumbrations and the Generation of the Three-Dimensional Space: Strategies in Embodiment and in Distributed Cognition Environment

This subsection (and the subsequent one) is devoted to study—in the light of abductive cognition—the so-called "anticipations": they will help us to delineate both the role of strategies in distributed hypothetical reasoning and to further illustrate the main features of the so-called "unlocked" strategies. Indeed, as I have already indicated, when describing manipulative abduction, strategic cognition not only refers to propositional aspects but it is also performed in a distributed cognition framework, in which models, artifacts, internal and external representations, sensations, and manipulations play an important role: indeed the phenomenological example illustrated in this subsection also shows that strategic cognition can involve, when clearly seen in embodied and distributed systems, visual, kinesthetic, and motor sensations. In this case we deal with a "natural game" between humans and their surroundings, in which "unlocked" strategies are at play: in the light of the analysis of the distinction between unlocked and locked strategies it will be useful to compare this case with the

[9] AlphaGo Zero is a version of DeepMind's Go software AlphaGo; the recent AlphaZero, that learns from games autonomously played, further enriches AlphaGo Zero, cf. https://en.wikipedia.org/wiki/AlphaZero.

human made case of "artificial games", to show instructive analogies and differences (see Sects. 3.2.3 and 3.3).

Looking at the philosophical explanations of the ways humans perform to build "idealities", geometrical idealities, and the objective space, Husserl contends that "facticities of every type [...] have a root in the essential structure of what is generally human", and that "human surrounding world is the same today and always" (Husserl (1978, p. 180)). Of course this should not hold when we consider the possible evolutionary character of this surrounding world and so of organic beings. A similar kind of possibility was advanced by Helmholtz and Poincaré, when they hypothesized the famous "fantastic worlds" in which there are beings educated in an environtment quite different from ours (Poincaré (1902, pp. 64–68)). Their different "experience" will lead these beings to classify phenomena in a different way than we would, that is a non-Euclidean way, because it is more convenient, even though the same phenomena could be described in a Euclidean way. In fact, Poincaré says that these worlds can be described "without forsaking the use of ordinary geometrical language" (1902, p. 71). To the aim of my considerations in this chapter, which regards "beings like us" this objection can be disregarded.

However, the horizon of the rough surrounding prepredicative world of appearances and primordial and immediately given experiences—which is at the basis of the constructive cognitive activity—is a source of potentially infinite data,[10] which cannot "lock" cognitive strategies related to the multiple strategic abductive generation of idealities, geometrical ideal forms, and spatiality. Indeed, step by step, ideal objects in Husserlian sense are constructed and become *traditional* objects, and so they possess historicity as one of their multiple eidetic components. They become, Husserl says, "sedimentations of a truth meaning", which describe the cumulative character of human experience (not every "abiding possession" of mine is traceable to a self-evidence of my own). The research which takes advantage of the already available sedimented idealities (sedimentations of someone else's already accomplished experience) is at the basis of further abductive work to the aim, for example, of discovering new mathematical knowledge in the field of geometry.

Let us follow some Husserlian speculations that lead us to consider the important strategic role of anticipations as abductions. In a recent article (Magnani 2015b) I have illustrated the constitutive abductive character of perception in the light of Peircean philosophy. Now we will see the strategic abductive role of both perception and kinesthetic data in the Husserlian philosophical framework, integrating it with a reference to some of the current results of neuroscience. Indeed, the philosophical tradition of phenomenology fully recognizes the protogeometrical role of kinesthetic data in the generation of the so-called "idealities" (and of geometrical idealities). The objective space we usually subjectively experience has to be put in brackets by means of the transcendental reduction, so that pure lived experiences can be examined without the compromising intervention of any psychological perspective, any "doxa". By means of this transcendental reduction, we will be able to recognize perception

[10] The prepredicative world is not yet characterized by predications, values, empirical manipulations, and techniques of measurement as instead the Husserl's prescientific world is.

as a structured "intentional constitution" of the external objects, established by the rule-governed activity of consciousness (similarly, space and geometrical idealities, like the Euclidean ones, are "constituted" objective properties of these transcendental objects).

The modality of appearing in perception is already markedly structured: it is not that of concrete material things immediately given, but it is mediated by sensible schemata constituted in the temporal continual mutation of adumbrations. So at the level of "presentational perception" of pure lived experiences, only partial aspects (*adumbrations* [*Abschattungen*]) of the objects are provided. Therefore, an activity of unification of the different adumbrations to establish they belong to a particular and single object (noema) is further needed.[11]

The analysis of the generation of idealities (and geometrical idealities) is constructed in a very thoughtful philosophical scenario. The noematic appearances are the objects as they are intuitively and immediately given (by direct acquaintance) in the constituting multiplicity of the so-called adumbrations, endowed with a morphological character. The noematic meaning consists of a syntactically structured categorical content associated with judgment. Its ideality is "logical". The noema consists of the object as deriving from a constitutive rule or synthetic unity of the appearances, in the transcendental sense (Petitot 1999). To further use the complex Husserlian philosophical terminology—which surely motivates an interpretation in terms of abduction—we can say: hyletic data (that is immediate given data) are vivified by an intentional synthesis (a noetic apprehension) that transforms them into noematic appearances that adumbrate objects, etc.

As illustrated by Husserl in *Ding und Raum* (Husserl 1973) the geometrical concepts of point, line, surface, plane, figure, size, etc., used in eidetic descriptions are not spatial "in the thing-like sense": rather, in this case, we deal with the problem of the generation of the objective space itself. Husserl observes: it is "senseless" to believe that "the visual field is [...] in any way a surface on objective space" (Sect. 48, p. 166), that is, to act "as if the oculomotor field were located, as a surface, in the space of things" (Sect. 67, p. 236). Moreover, Husserl thinks that space is endowed with a double function: it is able to constitute a phenomenal extension at the level of sensible data and also furnishes an intentional moment. Petitot says: "Space possesses, therefore, a noetic face (format of passive synthesis) and a noematic one (pure intuition in Kant's sense)" (1999, p. 336). What about the phenomenological genesis of geometrical global three-dimensional space?

The process of making adumbrations represents a strategy that is distributed in visual, kinesthetic, and motor activities usually involving the manipulations of some parts of the external world. The adumbrative aspects of things are part of the visual field. To manage them a first requirement is related to the need of gluing different fillings-in of the visual field to construct the temporal continuum of perceptive

[11] On the role of adumbrations in the genesis of ideal space and on their abductive and nonmonotonic character cf. Sect. 3.2.3. An interesting article (Overgaard and Grünbaum 2007) deals with the relationship between perceptual intentionality, agency, and bodily movement and acknowledges the abductive role of adumbrations. In the remaining part of this section I will try to clarify their meaning.

adumbrations in a global space: the visual field is considered not translation-invariant, because the images situated at its periphery are less differentiated than those situated at its center (and so resolution is weaker at the periphery than at the center), as subsequently proved by the pyramidal algorithms in neurophysiology of vision research.

Perceptual intentionality basically depends on the ability to realize kinesthetic situations and sequences. In order for the subject to have visual sensations of the world, he/she must be able not only to possess kinesthetic sensations but also to freely initiate kinesthetic strategic "sequences": this involves a bodily sense of agency and awareness on the part of the doer (Overgaard and Grünbaum, 2007, p. 20). The kinesthetic control of perception is related to the problem of generating the objective notion of three-dimensional space, that is, to the phenomenological constitution of a "thing",[12] as a single body unified through the multiplicity of its appearances. The "meaning identity" of a thing is of course related to the continuous flow of adumbrations: given the fact that the incompleteness of adumbrations implies their synthetic consideration in a temporal way, the synthesis in this case, *kinetic*, involves eyes, body, and objects.

Visual sensations are not sufficient to constitute objective spatiality. Kinesthetic sensations[13] (relative to the movements of the perceiver's own body)[14] are required. Petitot observes, de facto illustrating the abductive role of kinesthetic sensations:

> Besides their "objectivizing" function, kinesthetic sensations share a "subjectivizing" function that lets the lived body appear as a proprioceptive embodiment of pure experiences, and the adumbrations as subjective events. [...] There exists an obvious equivalence between a situation where the eyes move and the objects in the visual field remain at rest, and the reciprocal situation where the eyes remain at rest and the objects move. But this trivial aspect of the relativity principle is by no means phenomenologically trivial, at least if one does not confuse what is constituting and what is constituted. Relativity presupposes an *already* constituted space. At the preempirical constituting level, one must be able to discriminate the two equivalent situations. The kinesthetic control paths are essential for achieving such a task (Petitot (1999, pp. 354–355)).

Multidimensional and hierarchically organized, the space of kinesthetic controls includes several degrees of freedom for movements of eyes, head, and body. Kinesthetic controls are kinds of *spatial* gluing operators. They are able to compose, in the case of visual field, different partial aspects—identifying them as belonging to the same entity, (cf. Fig. 3.1), that is constituting an ideal and transcendent "object". They are realized in the pure consciousness and are characterized by an intentionality that demands a temporal lapse of time.

With the help of very complex eidetic descriptions, that further develop the strategic operations we sketched, Husserl is able to explain the constitution of the objective

[12] Cf. also (Husserl, 1931, Sect. 40, p. 129) [originally published in 1913].

[13] Husserl uses the terms "kinestetic sensations" and "kinesthetic sequences" to denote the subjective awareness of position and movement in order to distinguish it from the position and movement of perceived objects in space. On some results of neuroscience that corroborate and improve several phenomenological intuitions cf. Pachoud (1999, pp. 211–216), Barbaras (1999), and Petit (1999).

[14] The ego itself is only constituted thanks to the capabilities of movement and action.

Fig. 3.1 Scanning a cat with a partial image of it (very simple and unequivocal adumbration). To the left part corresponds a token of the whole visual field *D* centered on *p* (focalization on *p*)

parametrized time and of space, dealing with stereopsis, three-dimensional space and three-dimensional things inside it. Of course, when the three-dimensional space (still inexact) is generated (by means of two-dimensional gluing and stereopsis) it is possible to invert the phenomenological order: the visual field is so viewed as a portion of surface in $\mathbf{R}^3$, and the objective constituted space comes first, instead of the objects as they are intuitively and immediately given by direct acquaintance. So the space is in this case an objective datum informing the cognitive agent about the external world where she can find objects from the point of view of their referentiality and denotation. The kinesthetic system "makes the oculomotor field (eventually enlarged to infinity) the mere projection of a three spatial thingness" (Husserl, 1973, Sect. 63, p. 227). Adumbrations now also appear to be consequences of the objective three-dimensional space, as continuous transformations of two-dimensional images as if the body were embedded in the space $\mathbf{R}^3$. The role of adumbrations in objectifying entities can be also hypothesized in many cases of nonlinguistic animal cognition dealing with the problem of reification and the formation of a kind of "concept", cf. chapter five of (Magnani 2009). In human adults objects are further individuated and reidentified by using both spatial aspects, such as place and trajectory information and static-property information (in this last case exploiting what was gained through previous adumbration activity); adults use this property information to explain and predict appearances and disappearances "If the same large, distinctive white rabbit appears in the box and later on in the hat, I assume it's the same rabbit" (Gopnik and Meltzoff 1997).

Let us abandon the problem of the prelinguistic cognitive abductive strategies which are at play in a natural interaction—natural game—between humans and their prepredicative surroundings and let us concentrate on the cognitive abductive strategies that are at play in the artifactual case of the moves that are occurring in the adversarial game Go with two players and their respective changing surroundings, which in this case are basically formed by board, stones, and possible artifactual assisting accessories. In this game, analogously to the case of the natural processes, we still obviously find the role of visual, kinesthetic, and motor sensations and actions, but also the strong function of visual, iconic, and propositional artificial representations, anchored to the human made "meanings" (both internal and external) which gave birth to the game and which characterize its features and rules.

3.2.3 Anticipations as Abductions

Of course adumbrations, the substrate of gluing operations that give rise to the two-dimensional space, are multiple and infinite, and there is a potential co-givenness of some of them (those potentially related to single objects). They are incomplete and partial so for the complete givenness of an object a temporal process is necessary. Adumbrations, not only intuitively presented, can be also represented at the level of *imagination* and of course are kinds of anticipation. An *anticipation* is of course incomplete, and corresponds to a kind of non-intuitive intentional expectation: when we see a spherical form from one perspective (as an adumbration), we will assume that it is effectively a sphere, but it could be also a hemisphere (an example already employed by Locke).

Anticipations share with visual and manipulative abduction various features: they are highly conjectural and nonmonotonic, so wrong anticipations have to be replaced by other plausible ones. Moreover, they constitute an activity of "generate and test" as a kind of action-based cognition: the finding of adumbrations involves kinesthetic controls, sometimes in turn involving manipulations of objects; but the activity of testing anticipations also implies kinesthetic controls and manipulations. Finally, not all the anticipations are informationally equivalent and work like attractors for privileged individuations of objects. In this sense the whole activity is toward "the best anticipation", the one that can display the object in an optimal way. Prototypical adumbrations work like structural-stable systems, in the sense that they can "vary inside some limits" without altering the apprehension of the object.

As in the case of selective abduction, anticipations are able to select possible paths for constituting objects, actualizing them among the many that remain completely tacit. As in the case of creative abduction, they can construct new ways of aggregating adumbrations, by delineating the constitution of new objects/things. In this case they originate interesting "attractors" that give rise to *new* "conceptual" generalizations.

Some of Husserl's wonderful philosophical speculations are being further developed scientifically from the neurological and cognitive perspective in current cognitive science research. Grush (2004b, 2007) has built an emulation theory based on control theory where forward models as emulators (shared by humans and many other animals) are used to illustrate, in the case of humans, various cognitive processes like perception, imagery, reasoning, and language. He contends that simulation circuits are able to hypothesize forward mapping from control signals to the anticipated—and so abduced—consequences of executing the control command. In other words, they mimic the body and its interaction with the environment, enhancing motor control through sensorimotor abductive hypotheticals: "For example, in goal-directed hand movements the brain has to plan parts of the movement before it starts. To achieve a smooth and accurate movement proprioceptive/kinesthetic (and sometimes visual) feedback is necessary, but sensory feedback per se is too slow to affect control appropriately" (Desmurget and Grafton 2002). The "solution" is an emulator/forward model that can predict the sensory feedback resulting from executing a particular motor command" (Svensson and Ziemke (2004, p. 1310)). The control theory framework is also useful to describe the emergence of implicit and explicit agency (Grush 2007). The humans' understanding of themselves as explicit agents is accomplished through an interplay between the standard egocentric point of view and the so-called "simulated alter-egocentric" point of view, which represents the agent itself as an entity in the environment.

Given the fact that motor imagery can be seen as the off-line driving force of the emulator via efference copies, it is noteworthy that the emulation theory can be usefully extended to account for visual imagery as the off-line operator behind an emulator of the motor-visual loop. In these systems a kind of *amodal* spatial imagery can be hypothesized: "Modal imagery [...] is imagery based on the operation of an emulator of the sensory system itself, whereas amodal imagery is based on the operation of an emulator of the organism and its environment: something like arrangements of solid objects and surfaces in egocentric space. I show how the two forms of emulation can work in tandem" (Grush, 2004b, p. 386).

It is important to note that amodal imagery is neither sentential nor pictorial because the amodal environment space/objects emulators are closely tied to the organism's sensorimotor engagement with the environment. An interesting example of amodal abduction, in our terms, "where an object cannot currently be sensed by any sensory modality (because it is behind an occluder, is silent and odorless, etc.) yet it is represented as being at a location. I think it is safe to say that our representation of our own behavioral (egocentric) space allows for this, and it is not clear how a multisensory system, in which tags for specific modalities were always present, could accomplish this" (Grush (2004a, p. 434)).[15]

The Husserlian phenomenological explanation of the generation of "idealities" leads to a moment in which, once the space as an objective datum is settled, it informs the cognitive agent about the external world where she can find objects from the point of view of their referentiality and denotation, like it is happening—currently—to beings like us.

[15] On Grush's approach cf. the detailed discussion illustrated in (Clark (2008, chapter seven)) in the framework of the theory of the extended mind; a treatment of current cognitive theories, such as the sensorimotor theory of perception, which implicitly furnish a scientific account of the phenomenological concept of anticipation, is given in chapter eight of the same book. A detailed treatment of recent neuroscience achievements which confirms the abductive character of perception is given in the article "Vision, thinking, and model-based inferences" (Raftopoulos 2017), recently published in the *Handbook of Model-Based Science* (Magnani and Bertolotti 2017).

3.2.4 Reading Ahead

A fundamental strategy we immediately detect in artificial games such as Go, which is necessary for proficient and smart tactical play, is the capacity to *read ahead*, as the Go players usually say. Reading ahead is a practice of generating groups of anticipations that aim at being very robust and complex (either serious-minded or intuitive) and that demand the consideration of

1. clusters of moves to be adopted and their potential outcomes. The available scenario at time t_1, exhibited by the board, represents an adumbration of a subsequent potential more profitable scenario at time t_2, which indeed is abductively credibly hypothesized: in turn, one more abduction is selected and actuated, which—consistently and believably—activates a particular move that can lead to an envisaged more fruitful scenario.
2. Possible countermoves to each move.
3. Further chances after each of those countermoves. It seems that some of the smarter players of the game can read up to 40 moves ahead even in hugely complex positions.

In a book published in Japan, related to the description of various strategies that can be exploited in Go games, Davies emphasizes the role of "reading ahead":

> The problems in this book are almost all reading problems. [...] they are going to ask you to work out sequences of moves that capture, cut, link up, make good shape, or accomplish some other clear tactical objective. A good player tries to read out such tactical problems in his head before he puts the stones on the board. He looks before he leaps. Frequently he does not leap at all; many of the sequences his reading uncovers are stored away for future reference, and in the end never carried out. This is especially true in a professional game, where the two hundred or so moves played are only the visible part of an iceberg of implied threats and possibilities, most of which stay submerged. You may try to approach the game at that level, or you may, like most of us, think your way from one move to the next as you play along, but in either case it is your reading ability more than anything else that determines your rank (Davies 1995 p. 6).

Further strategies that are usefully adopted by human players in the game Go are for instance related to "global influence, interaction between distant stones, keeping the whole board in mind during local fights, and other issues that involve the overall game. It is therefore possible to allow a tactical loss when it confers a strategic advantage".[16]

The material and external scenarios (which are composed by the sensible objects—stones and board) that characterize artificial games are the fruit of a cognition "sedimented"[17] in their embodiment, after the starting point of their creation and subsequent uses and modifications. The cognitive tools that are related to the application of both the game allowed rules and the individual inferential talents owned by the two

[16] Cf. Wikipedia, entry Go (game) https://en.wikipedia.org/wiki/Go_(game).

[17] An expressive adjective still used by Husserl (1978). Translated by D. Carr and originally published in *The Crisis of European Sciences and Transcendental Phenomenology* .

players, strategies, tactics, heuristics, etc. are sedimented in those material objects (artifacts, in this case) that become *cognitive mediators*:[18] for example they orient players' inferences, transfer information, and provoke reasoning chances. Once represented internally, the external subsequent scenarios become the object of mental manipulation and new ones are further made, to the aim of producing the next most successful move.

It is important to note again that these strategies, when actuated, are certainty characterized by an extended variety, but all are "locked", because the elements of each scenario are always the same (what changes is merely the number of seeable stones and their dispositions in the board), in a finite and stable framework (no new rules, no new objects, no new boards, etc.) These strategies are devoid of the following feature: they are not able to recur to reservoirs of information *different* from the ones available in the fixed given scenario. It is important to add a central remark: of course the "human" player can enrich and fecundate his strategies by referring to internal resources not necessarily directly related to the previous experience with Go, but with other preexistent skills belonging to disparate areas of cognition. This is the reason why we can say that the strategies of a "human" player present a less degree of closure with respect to the automatic player AlphaGo. In humans, strategies are locked with respect to the external rigid scenario, but more open with respect to the mental field of reference to previous wide strategic experiences; in AphaGo and in deep learning systems, the strategic reservoir cannot—at least currently—take advantage of that mental openness and flexibility typical of human beings: the repertoire is merely formed/learned to play the game by checking data of thousands of games, and no other sources.

I have also to say that the notion of cognitive locked strategy I am referring to here is not present in and it is unrelated to the usual technical categorizations of game theory. Fundamentally, in combinatorial game theory, Go can be technically illustrated as zero-sum (player choices do not increment resources available-colloquially), perfect-information, partisan, deterministic strategy game, belonging to the same class as chess, checkers (draughts) and Reversi (Othello). Moreover, Go is bounded (every game has to end with a victor within a finite number of moves and time), strategies are obviously associative (that is in function of board position), format is of course non-cooperative (no teams are allowed), positions are extensible (that is they can be represented by board position trees).[19]

---

[18] This expression, I have extendedly used in Magnani (2001), is derived from Hutchins, who introduced the expression "mediating structure", which regards external tools and props that can be constructed to cognitively enhance the activity of navigating. Written texts are trivial examples of a cognitive "mediating structure" with clear cognitive purposes, so mathematical symbols, simulations, and diagrams, which often become "epistemic mediators", because related to the production of scientific results: "Language, cultural knowledge, mental models, arithmetic procedures, and rules of logic are all mediating structures too. So are traffic lights, supermarkets layouts, and the contexts we arrange for one another's behavior. Mediating structures can be embodied in artifacts, in ideas, in systems of social interactions [...]" (Hutchins 1995 pp. 290–291) that function as an enormous new source of information and knowledge.

[19] Cf. Wikipedia entry Go (game) https://en.wikipedia.org/wiki/Go_(game).

3.3 Locked Abductive Strategies Counteract the Maximization of Eco-Cognitive Openness

As I have already anticipated above in Sect. 1.1.1, in my research I have recently emphasized (Magnani, 2017, Chap. 7) the *knowledge enhancing* character of abduction. This means that in this case the abductive reasoning strategies grant successful and highly creative outcomes. The knowledge enhancing feature regards several kinds of new generated knowledge of various novelty degrees, from that new knowledge about a suffering patient we have abductively accomplished in medical diagnosis (a case of selective abduction, as no new biomedical knowledge is created, just new knowledge about a person) to the new knowledge developed in scientific discovery, which many epistemologists celebrated, for example Paul Feyerabend in *Against Method* (1975). In the case of an artificial game such as Go, the knowledge activated thanks to an intelligent choice of already available strategies or thanks to the invention of novel strategies and/or heuristics must also be considered a result of knowledge enhancing abduction.

I strongly contend that, to arrive to uberous selective or creative optimal abductive results, useful strategies must be applied, but it is also needed to be in presence of a cognitive environment marked by what I have called *optimization of eco-cognitive situatedness*, in which eco-cognitive openness is fundamental (Magnani 2016). This feature of the cognitive environment is especially needed in the case of strong creative abduction, that is when the kind of novelty is not restricted to the case of the "simple" successful diagnosis. In Sect. 3.4, I will illustrate in more detail that, to favor good creative and selective abductive reasoning, strategies must not be "locked" in an external restricted eco-cognitive environment, such as in a scenario characterized by fixed defining rules and finite material aspects, which would function as cognitive mediators able to constrain agents' reasoning.

At this point, it is valuable to supply a brief presentation of the concept of *eco-cognitive openness*. The new viewpoint inaugurated by the so-called naturalization of logic (Magnani 2005a) contends that the regulating authority claimed by formal models of ideal reasoners to control human reasoning practice on the ground is, to date, unwarranted. It is instead urgent to propose a "naturalization" of the logic of human inference. Woods held a naturalized logic to an adequacy condition of "empirical sensitivity" (Woods 2013). A naturalized logic is open to considering many ways of thinking that are typical of concrete human knowers, such as fallacies, which, even if not truth preserving inferences, in any case can furnish truths and/or profitable results. Of course, one of the guiding cases is furnished by the logic of abduction, where the naturalization of the well-known fallacy "affirming the consequent" is at play. As already anticipated, in my recent research on abduction (Magnani 2015b, 2016, 2017), I emphasized the importance in good abductive cognition to hypotheses of what has been called *optimization of situatedness*.

Let us explain what is the meaning of the expression optimization of situatedness: abductive cognition is for example very important in scientific reasoning because it refers to that activity of creative hypothesis generation which characterizes one of the more valued aspects of rational knowledge. To get abductive results in science, the "situatedness" of the involved cognitive activities is strongly connected with eco-cognitive aspects, related to the contexts in which knowledge is "traveling": in the case of scientific abductive cognition (but also in other abductive cases, such as diagnosis) to favor the solution of an inferential process the situatedness also has to be characterized by the richness of the flux of information, which has to be maximized. This maximization aims at that optimization of situatedness, which, as I quoted above, can only be made possible by a *maximization of changeability* of the basic data which informs the abductive cognitive process: inputs have to be maximally enriched, rebuilt, or modified and the same has to occur with respect to the knowledge applied during the hypothetical reasoning process. Obviously, the aim is to have at disposal a favorable "cognitive environment" in which available data can become *optimally positioned*.[20]

In summary, abductive processes to hypotheses—in a considerable quantity of cases, for example in science—are highly *information-sensitive*, and face with a flow of information and data uninterrupted and appropriately promoted and enhanced when needed (of course also thanks to artifacts of various kinds). This means that also from the psychological perspectives of the individuals the epistemological openness in which knowledge is channeling has to be maximized is fundamental.

A note on the history of philosophy can be added: already Aristotle provided a first fundamental study on abduction, which stresses the relevance, we can hazard, of *non-locked*, but highly open, cognition, in the celebrated (by Charles Sanders Peirce) passage of Chapter B25 of *Prior Analytics* regarding ἀπαγωγή (that is abduction, translated, in the English edition, with "leading away"). Indeed, it is exactly the idea of "leading away" which expresses that in smart abductions we have to integrate (or "unlock") the given components of the cognitive environment with the help of other cognitive tools and data that are away from them.

3.4 Locking Strategies Restricts Creativity

Optimization of situatedness is related to unlocked strategies. Locked strategies, such as the ones active in Go game, AlphaGo, and other computational AI systems and deep learning devices, do not favor the optimization of situatedness. Indeed, I have already contended above that, to obtain good creative and selective abductions, reasoning strategies must not be "locked" in bounded eco-cognitive surroundings (that is, in scenarios designed by fixed defining rules and finite material objects which would play the role of the so-called cognitive mediators). In this perspective, a poor scenario is certainly responsible for the minimization of the eco-cognitive openness and it is the structural consequence of the constitutive organization of the game Go (and also of Chess and other games), as I have already described in Sect. 3.2.4. I have said that in the game Go stones, board, and rules are rigid

[20] I have furnished more cognitive and technical details to explain this result in (Magnani 2016).

and so totally predetermined; what instead is undetermined are the strategies and connected heuristics that are adopted to defeat the adversary in their whole process of application.[21]

However, as I have already said, the available strategies and the adversary's ones are always *locked* in the fixed scenario: you cannot, during a Go game, play for a few minutes Chess or adopt another rule or another unrelated cognitive process, affirming that that weird part of the game is still appropriate to the game you agreed to play. You cannot decide to change the environment at will so *unlocking* your strategic reasoning, for example because you think this will be an optimal way to defeat the adversary. Furthermore, your adversary cannot activate at his discretion a process of eco-cognitive transformation of that artificial game. On the contrary, in the example of scientific discovery, the scientist (or the community of scientists) frequently recur to disparate external models and change their reasoning strategies[22] to produce new analogies or to favor other cognitive useful procedures (prediction, simplification, confirmation, etc.) to enhance the abductive creative process.

The case of scenarios in human scientific discovery precisely represents the counterpart of the ones that are poor from the perspective of their eco-cognitive openness. Indeed, in these last cases, the reasoning strategies that can be endorsed (and also created for the first time), even if multiple and potentially infinite, are *locked* in a determined perspective where the components do not change (the stones can just diminish and put aside, the board does not change, etc.) I would say that in scenarios in which strategies are locked, in the sense I have explained, an *autoimmunization* (Magnani and Bertolotti 2011; Arfini and Magnani 2015) is active, that constitutes the limitations that preclude the application of strategies that are not related to "prepackaged" scenarios, strategies that would be foreigners to the ones that are strictly intertwined with the components of the given scenario. Remember I already said that these components play the role of *cognitive mediators*, which anchor and constrain the whole cognitive process of the game.

To summarize and further explain (by linking the problem of locked and unlocked strategies to the various cases of selective and creative abduction):

1. Contrarily to the case of high level "human" creative abductive inferences such as the ones expressed by scientific discovery or other examples of special exceptional intellectual results, the status of artificial games (and of their computational counterpart) is very poor from the point of view of the non-strategic knowledge involved. We are dealing with stones, a modest number of rules, and one board. When the game progresses, the shape of the scenario is spectacularly modified but no unexpected cognitive mediators (objects) are appearing: for example, no diversely colored stones, or a strange hexagonal board. On the contrary, to continue with the example of high levels creative abductions in scientific discovery (for example, in empirical science), first of all the evidence is extremely rich and endowed with often unexpected novel features (not only due to modifications of

[21] Of course, many of the strategies of a good player are already mentally present thanks to the experience of several previous games.

[22] Many interesting examples can be found in the recent (Magnani and Bertolotti 2017).

aspects of the "same things", as in the case of artificial games). Secondly, the flux of knowledge at play is multifarious and is related to new analogies, thought experiments, models, imageries, mathematical structures, etc. that are rooted in heterogeneous disciplines and fields of intellectual research. In sum, in this exemplary case, we are facing a real tendency to a status of optimal eco-cognitive situatedness (further details on this kind of creative abduction are furnished in (Magnani 2015b, 2016, 2017).

2. What happens when we are dealing with selective abduction (for example in medical diagnosis)? First of all, evidence freely and richly arrives from several empirical sources in terms of body symptoms and data mediated by sophisticated artifacts (which also change and improve thanks to new technological inventions). Second, the encyclopedia of biomedical hypotheses in which selective abduction can work is instead locked,[23] but the reference to possible new knowledge (locally created of externally available) is not prohibited, so the diagnostic inferences can be enhanced thanks to scientific advancements at a first sight not considered. Third, novel inferential strategies and linked heuristics can be created and old ones used in new surprising ways but, what is important, strategies are not locked in a fixed scenario. In sum, the creativity that is occurring in the case of human selective abduction is poorer than the one active in scientific discovery, but richer than the one related to the activity of the locked reasoning strategies of the Go game and AlphaGo, I have considered above.
3. In Go (and similar games) and in deep learning systems such as AlphaGo, in which strategies and heuristics are "locked", these are exactly the only part of the game that can be improved and rendered more fertile: strategies and related heuristics can be used in a novel way and new ones can be invented. Anticipations as abductions (which incarnate the activities of "reading ahead") just affect the modifications and re-grouping of the same elements. No other types of knowledge will increase; all the rest remains stable.[24] Of course, this dominance of the strategies is the quintessence of Go, Chess, and other games, and also reflects the spectacularity of the more expert moves of the human champions. However, it has to be said that this dominance is also the reason that explains the fact the creativity at stake is even more modest than the one involved in the higher cases of selective abduction (diagnosis). I will soon illustrate that this fact is also the reason that explains why the smart strategies of Go or Chess games can be more easily simulated, for example with respect to the inferences at play in scientific discovery, by recent artificial intelligence programs, such as the ones based on deep learning.[25]

The reader does not have to misunderstand me: I do not mean to minimize the relevance of creative heuristics as they work in Go and other board games. John

[23] It is necessary to select from pre-stored diagnostic hypotheses.

[24] Obviously, for example, new rules and new boards can be proposed, so realizing new types of game, but this chance does not jeopardize my argumentation.

[25] Some notes on the area of the so-called automated scientific discovery in AI cf. (Magnani (2009, Chap. 2, Sect. 2.7 "Automatic Abductive Scientists")).

Holland already clearly illustrated (1995, 1997) that board games such as checkers, as well as Go, are wonderful cases of "emerging" cognitive processes, where potentially infinite strategies favor exceptional games: even if simply thanks to a few rules regulating the moves of the pieces, games cannot be predicted starting from the initial configurations. While other cases of emerging cognitive processes (I have indicated the example of scientific discovery) characterize what can be called "vertical" creativity (that is, related to unlocked strategies), board games are examples of "horizontal" creativity: even if board games are circumscribed by locked strategies that constrain the game, "horizontal" creativity can show astonishing levels of creativity and skillfulness. We already said that these extraordinary humans skills have been notably appropriated by artificial intelligence software (see below the last paragraphs of this section): the example given in this chapter is the one of AI deep learning heuristics that were able to *learn from* human games, such in the case of AlphaGo, or from games autonomously played, such in the case of AlphaZero, as I have already said. What are the remaining most important effects which derive from these computational AI programs equipped to concretize cognitive abductive inferences characterized by "locked" strategies?

At the beginning of this chapter, I have illustrated some amazing performances of Google DeepMind's program AlphaGo against human players and the fact the system showed to be able to "create" unconventional moves—never played by humans—thus building new strategies, a fact that undoubtedly favors the attribution to the system of actual "human" capacities, and we have to add, better than the ones of the more skilled humans. AlphaGo instructed itself to play "attending" (so to speak) thousands of games played by human beings thanks to "reinforcement learning", which refers to that activity of self-playing of the machine (the machine plays against itself) to further feed and adapt its own neural networks.

In the passage I am reporting below, Cohleo and Thompsen Primo seem to corroborate my contention regarding the fact I consider AI programs as AlphaGo substantially simple to be produced at the computational level because they are dealing with what I have called in this chapter *locked reasoning strategies*. In summary, a kind of background reason for this easiness would be that this type of human cognition is less creative than others, even if it is so striking when optimally realized by very smart human intelligent subjects.

> Let us compare the key ideas behind Deep Blue (Chess) and AlphaGo (Go). The first program used values to assess potential moves, a function that incorporated lots of detailed chess knowledge to evaluate any given board position and immense computing power (brute force) to calculate lots of possible positions, selecting the move that would drive the best possible final possible position. Such ideas were not suitable for Go. A good program may capture elements of human intuition to evaluate board positions with good shape, an idea able to attain far-reaching consequences. After essays with Monte Carlo tree search algorithms, the bright idea was to find patterns in a high quantity of games (150,000) with deep learning based upon neural networks. The program kept making adjustments to the parameters in the model, trying to find a way to do tiny improvements in its play. And, this shift was a way out to create a policy network through billions of settings, i.e., a valuation system that captures intuition about the value of different board positions. Such search-and-optimization idea was cleverer about how search is done, but the replication of intuitive pattern recognition

> was a big deal. The program learned to recognize good patterns of play leading to higher scores, and when that happened it reinforces the creative behavior (it acquired an ability to recognize images with similar style) (Coelho and Thompsen 2017, p. 21).

I think humans with their biological brains do not have to feel mortified by these extraordinary skillful capacities of the AI programs. Unfortunately, given the present worldwide status of mass media, other magnificent human performances in various fields of creativity, much more creative than the ones related to locked strategic reasoning, are unable to reach the global echo AlphaGo gained. Indeed, human-more-skillful-abductive creative capacities, related to unlocked strategies, as I have tried to demonstrate in this chapter—still cognitively beautiful—are not sponsored by Google, which is a herculean corporation that can easily obtain the attention of not only the monocultural media of our age, but also of the social networks: many human beings are more easily impressionable by the "miracles" of AI, robotics, and information technologies, than by prodigious knowledge results of human beings-like-us, too often out of sight (after all—ça va sans dire—also AI traditional programs and AI deep learning systems have been created by humans…)

Google managers also think that AI deep learning programs similar to AlphaGo could be exploited to help science resolve important real-world problems in health-care but also in other fields. This would be a good research program. Google seems to also expect to implement some business thanks to commercialization of new deep learning AI powers to collect appropriate information and generate abductions in some advantageous fields. Simply checking the Wikipedia entry DeepMind (https://en.wikipedia.org/wiki/DeepMind),[26] [DeepMind is a British artificial intelligence company instituted in September 2010 and took by Google in 2014, the company created the AlphaGo program], the following non-contested passage is reported, concerning the so-called "NHS data-sharing controversy":

> In April 2016, *New Scientist* obtained a copy of a data-sharing agreement between Deep-Mind and the Royal Free London NHS Foundation Trust. The latter operates in three London hospitals where an estimated 1.6 million patients are treated annually. The agreement shows DeepMind Health had access to admissions, discharge and transfer data, accident and emergency, pathology and radiology, and critical care at these hospitals. This included personal details such as whether patients had been diagnosed with HIV, suffered from depression or had ever undergone an abortion in order to conduct research to seek better outcomes in various health conditions. A complaint was filed to the Information Commissioner's Office (ICO), arguing that the data should be pseudonymised and encrypted. In May 2016, *New Scientist* published a further article claiming that the project had failed to secure approval from the Confidentiality Advisory Group of the Medicines and Healthcare Products Regulatory Agency. In May 2017, *Sky News* published a leaked letter from the National Data Guardian, Dame Fiona Caldicott, revealing that in her "considered opinion" the data-sharing agreement between DeepMind and the Royal Free took place on an "inappropriate legal basis". The Information Commissioner's Office ruled in July 2017 that the Royal Free hospital failed to comply with the Data Protection Act when it handed over personal data of 1.6 million patients to DeepMind.

Even if based on what I called in this chapter locked strategies, and thus far from the highest levels of human creativity, AI deep learning system and various other pro-

[26] Date of access August 19th, 2021.

grams can also offer chances for business and good integration in the market. I think epistemologists and logicians have to monitor the use of these AI devices (of course, when less transparent than the natural and limpid —and so stupefying—performance of AlphaGo in games against humans). Recent research in the field of epistemology, cognitive science, and philosophy of technology[27] illustrate that good AI software, which surely furnishes a big new chance for opportunity and data analytics, can be transmuted in a tool that does not respect epistemological and/or ethical rigor. For example, in the case regarding the computational exploitation of big data, outcomes can inadvertently lead to epistemologically unacceptable computer-discovered correlations (instead possibly good from a commercial perspective), but these tools are sometimes—unfortunately—seriously illustrated as aiming at replacing tout-court human-based scientific research as a guide to understanding, prediction, and action. Calude and Longo say: "Consequently, there will be no need to give scientific meaning to phenomena, by proposing, say, causal relations, since regularities in very large databases are enough: 'with enough data, the numbers speak for themselves' " (Calude and Longo 2017 p. 595). Unfortunately, some "correlations appear only due to the size, not the nature, of data. In 'randomly' generated, large enough databases too much information tends to behave like very little information" (*ibid.*). I agree with these authors: we cannot treat some spurious correlations as results of deep scientific creative abduction, but just as trivial generalizations, even if reached with the help of sophisticated artifacts.[28] I cannot further deepen the problems regarding issues connected to the impact of computational programs on ethics and society. In this chapter, I limit myself to deal with cognitive, logical, and epistemological aspects to the aim of introducing the distinction between locked and unlocked strategies and its meaning with respect to intelligent computation.

Summary

In this chapter, with the help of the concepts of locked and unlocked strategies, abduction, and optimization of eco-cognitive openness, I have described some central aspects of the cognitive character of reasoning strategies and related heuristics, to the aim of shedding new light on the cognitive aspects of deep learning machines. Taking advantage of my studies on abduction, I have contended that what I call *eco-cognitive openness* is undermined in the case of famous computational programs such as AlphaGo/AlphaZero, because they are based on *locked abductive strategies*. Instead, unlocked abductive strategies, which are in tune with what eco-cognitive openness requires, qualify those high-level kinds of abductive creative reasoning that are typical of human-based cognition. Locked abductive reasoning strategies are much simpler than unlocked ones to be rendered at the computational level: they indeed take advantage of a kind of autoimmunity that grants the limitations that preclude the application of strategies that are not related to "pre-packaged"

---

[27] Relatively recent bibliographic references can be found in my book (Magnani 2007).

[28] On this problem and other negative epistemological use of computational programs, cf. the recent (Calude and Longo 2017).

scenarios, strategies that would be foreign to the ones that are strictly intertwined with the components of a given scenario.

References

Arfini, S., Magnani, L.: An eco-cognitive model of ignorance immunization. In: Magnani, L., Li, P., Park, W. (eds.) Philosophy and Cognitive Science II. Western & Eastern Studies, vol. 20, pp. 59–75. Springer, Switzerland (2015)

Barbaras, R.: The movement of the living as the originary foundation of perceptual intentionality. In: Petitot, J., Varela, F.J., Pachoud, B., Roy, J.-M. (eds.) Naturalizing Phenomenology, pp. 525–538. Stanford University Press, Stanford, CA (1999)

Boden, M.: The Creative Mind: Myths and Mechanisms. Basic Books, New York (1992)

Boneh, D., Dunworth, C., Lipton, R.J., Sgall, J.: On the computational power of DNA, Discrete Applied Mathematics. Comput. Mol. Biol. **71**, 79–94 (1996)

Bruza, P.D., Cole, R.J., Song, D., Bari, Z.: Towards operational abduction from a cognitive perspective. Logic J. IGPL **14**(2), 161–179 (2006)

Bruza, P.D., Kitto, K., Ramm, B., Sitbon, L., Blomberg, S., Song, D.: Quantum-like non-separability of concept combinations, emergent associates and abduction. Logic J. IGPL **20**(2), 445–457 (2012)

Bylander, T., Allemang, D., Tanner, M.C., Josephson, J.R.: The computational complexity of abduction. Artif. Intell. **49**, 25–60 (1991)

Calude, C.S., Longo, G.: The deluge of spurious correlations in big data. Found. Sci. **22**(3), 595–612 (2017)

Clark, A.: Supersizing the Mind. Embodiment, Action, and Cognitive Extension. Oxford University Press, Oxford/New York (2008)

Coelho, H., Thompsen Primo, T.: Exploratory apprenticeship in the digital age with AI tools. Prog. Artif. Intell. **1**(6), 17–25 (2017)

Colton, S. (ed.): AI and Scientific Creativity. Proceedings of the AISB99 Symposium on Scientific Creativity, Society for the Study of Artificial Intelligence and Simulation of Behaviour. University of Edinburgh, Edinburgh, Edinburgh College of Art and Division of Informatics (1999)

Corruble, V., Ganascia, J.-G.: Induction and the discovery of the causes of scurvy: A computational reconstruction. Artif. Intell. **91**, 205–223 (1997)

Davies, J.: Tesuji. Elementary Go Series. 3. Tokyo: Kiseido Publishing Company (1995)

Davies, J., Goel, A.K.: A Computational Theory of Visual Analogical Transfer. Technical Report, Georgia Institute of Technology, Atlanta, GA (2000)

Davies, J., Goel, A.K., Yaner, P.W.: Proteus: Visual analogy in problem solving. Knowl.-Based Syst. **21**(7), 636–654 (2008)

de Kleer, J., Williams, B.: Diagnosing multiple faults. Artif. Intell. **32**, 97–130 (1987)

Desmurget, M., Grafton, S.: Forward modeling allows feedback control for fast reaching movements. Trends Cognit. Sci. **4**, 423–431 (2002)

Fajtlowicz, S.: On conjectures of Graffiti. Discrete Math. **72**, 113–118 (1988)

Falkenhainer, B.C.: A unified approach to explanation and theory formation. In: Shrager, J., Langley, P. (eds.) Computational Models of Scientific Discovery and Theory Formation, pp. 157–196. Morgan Kaufmann, San Mateo, CA. (1990)

Feyerabend, P.: Against Method. Verso, London-New York (1975)

Figueroa, A.R.: Inferencia abductiva basada en modelos. Una relación entre lógica y cognitión. Crítica. Revista Hispanoamericana de Filosofía **43**(129), 3–29 (2012)

Gärdenfors, P.: Conceptual Spaces: The Geometry of Thought. The MIT Press, Cambridge (2000)

Gigerenzer, G., Brighton, H.: Homo heuristicus: Why biased minds make better inferences. Topics Cognit. Sci. **1**, 107–143 (2009)

Gigerenzer, G., Selten, R.: Bounded Rationality. The Adaptive Toolbox. The MIT Press, Cambridge, MA (2002)
Glymour, C., Scheines, R., Spirtes, P., Kelly, K.: Discovering Causal Structure. Academic Press, San Diego, CA. (1987)
Gopnik, A., Meltzoff, A.: Words, Thoughts and Theories (Learning, Development, and Conceptual Change). The MIT Press, Cambridge, MA (1997)
Grush, R.: Agency, emulation and other minds. Cognit. Semiotics **0**, 49–67 (2007)
Grush, R.: Further explorations of the empirical and theoretical aspects of the emulation theory. Behavioral and Brain Sciences **27**, 425–435 (2004a). Author's Response to Open Peer Commentary to R. Grush, The emulation theory of representation: Motor control, imagery, and perception
Grush, R.: The emulation theory of representation: motor control, imagery, and perception. Behav. Brain Sci. **27**, 377–442 (2004b)
Holland, J.H.: Hidden Order. Addison-Wesley, Reading, MA (1995)
Holland, J.H.: Emergence: From Chaos to Order. Oxford University Press, Oxford (1997)
Husserl, E.: Ding und Raum: Vorlesungen (1907). Husserliana 16, edited by U. Claesges. Nijhoff, The Hague (1973)
Husserl, E.: Ideas. General Introduction to Pure Phenomenology [First book, 1913]. Translated by W. R. Boyce Gibson. London: Northwestern University Press (1931)
Husserl, E.: The Crisis of European Sciences and Transcendental Phenomenology [1954]. Translated by. D. Carr. George Allen & Unwin and Humanities Press, London (1970)
Husserl, E.: The Origin of Geometry (1939). In: Derrida, J. (ed.) Edmund Husserl's "The Origin of Geometry", pp. 157–180 (Translated by D. Carr and originally published in Husserl, 1970, pp. 353–378). Nicolas Hays, Stony Brooks, NY (1978)
Hutchins, E.: Cognition in the Wild. The MIT Press, Cambridge, MA (1995)
Klahr, D., Dunbar, K.: Dual space search during scientific reasoning. Cognit. Sci. **12**, 1–48 (1988)
Kulkarni, D., Simon, H.A.: The process of scientific discovery: The strategy of experimentation. Cognit. Sci. **12**, 139–176 (1988)
Langley, C., Klahr, D.: A 4-space model of scientific discovery. AAAI Symposium Systematic Methods of Scientific Discovery, Technical Report SS-95-03. AAAI Press, Menlo Park, CA (1995)
Langley, P., Simon, H.A., Bradshaw, G., Zytkow, J.: Scientific Discovery. Computational Explorations of the Creative Processes. The MIT Press, Cambridge, MA (1987)
Lenat, R.K.: Understanding diagrammatic demonstrations. In: Ram, A., Eiselt, K. (eds.) Proceedings of the 16th Annual Conference of the Cognitive Science Society, pp. 572–576. Erlbaum, Hillsdale, Paris (1994)
Lenat, D.: Discovery in mathematics as heuristic search. In: Davis, R., Lenat, D. (eds.) Knowledge-Based Systems in Artificial Intelligence. McGraw Hill, New York (1982)
Lindsay, R.K.: Using spatial semantics to discover and verify diagrammatic demonstrations of geometric propositions. In: O'Nuallian, S. (ed.) Spatial Cognition. Proceedings of the 16th Annual Conference of the Cognitive Science Society, pp. 199–212. John Benjamins, Amsterdam (2000b)
Lindsay, R.K.: Using diagrams to understand geometry. Comput. Intell. **9**(4), 343–345 (1998)
Lindsay, R.K.: Playing with diagrams. In: Anderson, M., Cheng, P., Haarslev, V. (eds.) Diagrams 2000, pp. 300–313. Springer, Berlin (2000a)
Lindsay, R.K., Buchanan, B., Feingenbaum, E., Lederberg, J.: Applications of Artificial Intelligence for Organic Chemistry: The Dendral Project. McGraw Hill, New York (1980)
Magnani, L.: Abduction, Reason, and Science. Processes of Discovery and Explanation. Kluwer Academic/Plenum Publishers, New York (2001)
Magnani, L.: Playing with anticipations as abductions. Strategic reasoning in an ecocognitive perspective. J. Appl. Logic - IfColog J. Logics and their Appl. **5**(5), 1061–1092 (2018). Special Issue on "Logical Foundations of Strategic Reasoning" (guest editors W. Park and J. Woods)
Magnani, L.: Morality in a Technological World. Knowledge as Duty. Cambridge University Press, Cambridge (2007)

Magnani, L.: Abductive Cognition. The Epistemological and Eco-Cognitive Dimensions of Hypothetical Reasoning. Springer, Heidelberg/Berlin (2009)
Magnani, L.: Naturalizing logic. errors of reasoning vindicated: Logic reapproaches cognitive science. J. Appl. Logic **13**, 13–36 (2015a)
Magnani, L.: The eco-cognitive model of abduction. Ἀπαγωγή now: Naturalizing the logic of abduction. J. Appl. Logic **13**, 285–315 (2015b)
Magnani, L.: The eco-cognitive model of abduction. Irrelevance and implausibility exculpated. J. Appl. Logic **15**, 94–129 (2016)
Magnani, L.: The Abductive Structure of Scientific Creativity. An Essay on the Ecology of Cognition. Springer, Cham, Switzerland (2017)
Magnani, L., Bertolotti, T.: Cognitive bubbles and firewalls: Epistemic immunizations in human reasoning. In: Carlson, L., Hölscher, C., Shipley, T. (eds.) CogSci 2011, XXXIII Annual Conference of the Cognitive Science Society. Cognitive Science Society, Boston MA (2011)
Magnani, L., Bertolotti, T. (eds.): Handbook of Model-Based Science. Springer, Switzerland (2017)
Newell, A., Shaw, J.C., Simon, H.A.: Empirical Explorations of the Logic Theory Machine: A Case Study in Heuristic. In: Proceedings of the Western Joint Computer Conference [JCC 11], pp. 218–239. Los Angeles February (1957)
Okada, T., Simon, H.A.: Collaborative discovery in a scientific domain. Cognit. Sci. **21**, 109–146 (1997)
O'Rorke, P., Morris, S., Schulemburg, D.: Theory formation by abduction: A case study based on the chemical revolution. In: Shrager, J., Langley, P. (eds.) Computational Models of Scientific Discovery and Theory Formation, pp. 197–224. Morgan Kaufmann, San Mateo, CA (1990)
Overgaard, S., Grünbaum, T.: What do weather watchers see? Perceptual intentionality and agency. Cognit. Semiotics **0**, 8–31 (2007)
Pachoud, B.: The teleological dimension of perceptual and motor intentionality. In: Petitot, J., Varela, F.J., Pachoud, B., Roy, J.-M. (eds.) Naturalizing Phenomenology, pp. 196–219. Stanford University Press, Stanford, CA (1999)
Paul, G.: AI approaches to abduction. In: Gabbay, D., Kruse, R. (eds.) Abductive Reasoning and Learning, pp. 35–98. Springer, Dordrecht (2000)
Pease, A., Colton, S., Smaill, A., Lee, J.: A model of Lakatos's philosophy of Mathematics. In: Magnani, L., Dossena, R. (eds.) Computing, Philosophy and Cognition, pp. 57–85. College Publications, London (2005)
Pennock, R.T.: Tower of Babel. The Evidence Against the New Creationism. The MIT Press, Cambridge, MA. (1999)
Pennock, R.T.: Can Darwinian mechanisms make novel discoveries? Learning from discoveries made by evolving neural network. Found. Sci. **5**(2), 225–238 (2000)
Pericliev, V., Valdés-Pérez, R.E.: Automatic componential analysis of kinship semantics with a proposed structural solution to the problem of multiple models. Anthropol. Linguist. **40**(2), 272–317 (1998)
Petit, J.-L.: Constitution by movement: Husserl in the light of recent neurobiological findings. In: Petitot, J., Varela, F.J., Pachoud, B., Roy, J.-M. (eds.) Naturalizing Phenomenology, pp. 220–244. Stanford University Press, Stanford, CA (1999)
Petitot, J.: Morphological eidetics for a phenomenology of perception. In: Petitot, J., Varela, F.J., Pachoud, B., Roy, J.-M. (eds.) Naturalizing Phenomenology, pp. 330–371. Stanford University Press, Stanford, CA (1999)
Poincaré, H.: La science et l'hypothèse (English translation by W. J. G. [only initials indicated], 1958, Science and Hypothesis, with a Preface by J. Larmor, The Walter Scott Publishing Co., New York, 1905. Also reprinted in Essential Writings of Henri Poincaré, Random House, New York, 2001). Flammarion, Paris (1902)
Raab, M., Gigerenzer, G.: Intelligence as smart heuristics. In: Sternberg, R.J., Prets, J.E. (eds.) Cognition and Intelligence. Identifying the Mechanisms of the Mind, pp. 188–207. Cambridge University Press, Cambridge, MA (2005)

Raftopoulos, A.: Vision, thinking, and model-based inferences. In: Magnani, L., Bertolotti, T. (eds.) Handbook of Model-Based Science, pp. 573–604. Springer, Switzerland (2017)
Rajamoney, S.A.: The design of discrimination experiments. Mach. Learn. **12**, 185–203 (1993)
Ramoni, M., Stefanelli, M., Magnani, L., Barosi, G.: An epistemological framework for medical knowledge-based systems. IEEE Trans. Syst., Man, Cybern. **22**(6), 1361–1375 (1992)
Reggia, J.A., Nau, D.S., Wang, P.Y.: Diagnostic expert systems based on set covering model. J. Man-Mach. Stud. **19**, 437–460 (1983)
Reiter, R.: A theory of diagnosis from first principles. Artif. Intell. **32**, 57–95 (1987)
Scott, P.D., Markovitch, S.: Experience selection and problem choice in an exploratory learning system. Mach. Learn. **12**, 49–67 (1993)
Shen, W.M.: Discovery as autonomous learning from the environment. Mach. Learn. **12**, 143–165 (1993)
Shrager, J., Langley, P. (eds.): Computational Models of Scientific Discovery and Theory Formation. Morgan Kaufmann, San Mateo, CA (1990)
Simon, H.A., Valdés-Pérez, R.E., Sleeman, D.H.: Scientific discovery and simplicity of method. Artif. Intell. **91**(2), 177–181 (1997)
Svensson, H., Ziemke, T.: Making sense of embodiment: Simulation theories and the sharing of neural circuitry between sensorimotor and cognitive processes. In: Forbus, K.D., Gentner, D., Regier, T. (eds.) CogSci 2004, XXVI Annual Conference of the Cognitive Science Society. CD-Rom, Chicago, IL (2004)
Swanson, D.R., Smalheiser, N.R.: An interactive system for finding complementary literatures: A stimulus to scientific discovery. Artif. Intell. **91**(2), 183–203 (1997)
Thagard, P.: Computational Philosophy of Science. The MIT Press, Cambridge, MA (1988)
Thagard, P.: Explanatory coherence. Behav. Brain Sci. **12**(3), 435–467 (1989)
Thagard, P.: Conceptual Revolutions. Princeton University Press, Princeton (1992)
Thagard, P., Litt, A.: Models of scientific explanation. In: Sun, R. (ed.) Cambridge Handbook of Computational Psychology, pp. 549–564. Cambridge University Press, Cambridge (2008)
Valdés-Pérez, R.E.: Principles of human computer collaboration for knowledge discovery in science. Artif. Intell. **107**(2), 335–346 (1999)
Woods, J.: Errors of Reasoning. Naturalizing the Logic of Inference. College Publications, London (2013)
Zeigarnik, A.V., Valdés-Pérez, R.E., Temkin, O.N., Bruk, L.G., Shalgunov, S.I.: Computer-aided mechanism elucidation of acetylene hydrocarboxylation to acrylic acid based on a novel union of empirical and formal methods. Organometallics **16**(14), 3114–3127 (1997)
Zytkow, J. (ed.): Proceedings of the ML-92 Workshop on Machine Discovery (MD-92). The Wichita State University, National Institute for Aviation Research (1992)

Chapter 4
Computational Domestication of Ignorant Entities

In the first two chapters of this book we have stressed that eco-cognitive computationalism sees computation in context, following some of the main tenets advanced by the recent cognitive science views on embodied, situated, and distributed cognition. We have also described the new attention in computer science devoted to the relevance in computation of the morphological features. It is by further deepening and analyzing the perspective opened by these novel fascinating approach that we see ignorant bodies as *domesticated* to become useful "mimetic bodies" from a computational point of view. This new perspective shows how the computational domestication of ignorant entities can originate new variegated unconventional cognitive embodiments, so opening the new research field of the so-called *natural computing*. In the last part of the chapter I will introduce the concept of overcomputationalism, showing that my proposed framework helps us see the related concepts of pancognitivism, paninformationalism, and pancomputationalism in a more naturalized and prudent perspective, avoiding the excess of old-fashioned ontological or metaphysical overstatements. Finally, I will submit to the attention of the reader a question that in my opinion synthesizes all the problems described in the chapter: will the future of eco-cognitive settings computationally-tailored or humanly-tailored.

4.1 Ignorant Physical and Biological Bodies

As described above in chapter two a recent area of study that concerns the so-called *morphological computation* deals with the problem of enriching classical digital computation taking advantage of "bodies", which are transformed into entities that carry their own computation in a useful (and unusual) way. It is in this framework of research that Hauser et al. (2014, p. 230) say that the body does not "know" that it is part of a computational device because it simply conforms to the laws of physics and it responds accordingly. Also, the body does not over- nor under-compensate,

L. Magnani, *Eco-Cognitive Computationalism*, Cognitive Systems Monographs 43,
https://doi.org/10.1007/978-3-030-81447-2_4

indeed it is a stable physical system. The proposed apparatus simply adds some linear readouts to the body to conclude the computation. The body would behave precisely the same, if there were no readouts at all. Moreover, the so-called Physical RC[1] is of practical value if it is able to provide a fruitful mapping of input streams onto output streams taking advantage of the physical body at hand. Finally, the authors introduce the role of ignorance: "Another interesting implication of this property of 'ignorance' of the body (i.e., it behaves the same, whether we add readouts or not) is that we can employ the same body for multiple computations on the same input at the same time. We simply have to add the according number of readouts" (Hauser et al. 2014, p. 230). An "absence of knowledge" is at play, the authors say that "the body still does not 'know' that it is part of this computational loop" and that this absence of knowledge "raises the philosophical question [...], whether the body is carrying out computation at all". The problem of *ignorance* explicitly emerges: the body "behaves the same, whether we add readouts or not" and "we can employ the same body for multiple computations on the same input at the same time. We simply have to add the according number of readouts".

Authors that incorporate in the so-called "computing nature" or "natural computation" the strong idea that whatever nature does can be framed in terms of computation as information processing will disagree with this attribution of ignorance to the body. They would say that, unlike inorganic matter, living cells are cognitive systems that possess agency and goals, and act in the world in predictive and adaptive way and that for a process to be computation it is not important to be part of computational mechanisms or even that whole computational machine knows that it is computing. They would also probably add that pebbles are certainly always ignorant, and so pebbles do not compute in the sense humans do, but that their very existence *is* a result of physical computational processes on a variety of levels of organization in nature. To avoid epistemological confusions I will further comment on this issue in the following subsection and below in Sect. 4.6, showing some puzzling problems related to the interpretation of the concept of natural computing.[2]

In this chapter I will discuss the problem of "ignorant entities" that carry computation as a fundamental recent issue of a philosophy of computing that does not aim at exhausting the concept of computation adopting the unilateral reduction to the digital definition (that is to the processing of strings of digits which agree with rules). We do not need a conclusive and rigid definition of the concept of computation, such as some mainstream epistemological approaches could provide: I contend that the concept has a historical and dynamical character.[3] Various types of computation have to be considered, which involve the change of the meaning of the concept thanks

[1] As I explained in chapter two of this book, the *Physical Reservoir Computing* (Physical RC) represents an interesting example of the novel perspective of morphological computation.

[2] In the following section I will treat both non-living and living bodies as "ideally" ignorant because I am interested in the "active" exploitation of both entities to the aim of getting conventional and unconventional computational performances endowed with cognitive consequences.

[3] Also Burgin and Dodig-Crnkovic (2015) present a rich taxonomy of existing notions of computation and propose to adopt an open-ended and evolving concept of computation that will be able to continuously absorb new insights from sciences.

to the emergence of new variances. In the present chapter, I will mainly, but not only, describe the consequences generated by the irruption in computation of the morphological chance.

I will take again advantage of what I call an *eco-cognitive perspective*, which leads to what I have described in the first two chapters of this book as "eco-cognitive computationalism" that is appropriate to clarify the meaning of the property of *ignorance* of the physical and biological bodies involved in a computation. Both perspectives will be necessary to illustrate, once and for all, the *historical* and *dynamical character* of the concept of *computation*: I do not aim at furnishing a final and stable definition of the concept of computation but an intellectual framework that depicts how we can understand not only the changes of its meaning, but also the "emergence" of new forms of computations, which precisely transform ignorant entities in cognitive entities through computation.

4.1.1 Different Kinds of Ignorance

The issue of the computational domestication of ignorant entities will be introduced in the next subsection. As for now, however, to avoid possible misunderstandings, it is useful to add some relevant observations regarding the different kinds of ignorance involved in physical and biological bodies. Indeed in this chapter, I am treating both non-living and living bodies as "ideally" ignorant because I am interested in the "active" exploitation of both entities to the aim of getting computational performances endowed with cognitive consequences. On the contrary, adopting a different intellectual perspective, which directly attributes cognitive capacities to living bodies and pays preeminent attention to this aspect, physical and biological entities present different kinds of ignorance. For example, from the rich and eclectic info-computational point of view proposed in Dodig-Crnkovic (2013, 2017), there is an important difference between nonliving and living bodies. Biological bodies possess cognition on different levels, from the lowest to the highest organization level. Even a single cell (the simplest living organism and the basic block of all living organisms) can be considered a cognitive system. A living cell in its *Umwelt* is not ignorant but a well-informed, active agent with intentions. Of course our cells in the body do not know what we are doing, but they "know"/cognize what they are doing. And so is the case with all living cells, they are informed by the environment, they learn, adapt, change their bodily structures, evolve. In their microscopic world they are alert, they can move towards the source of food, swim away from poison, using the environment for their own survival.[4]

Moreover, it can be hypothesized that in many organisms the perceptual world is the only possible model of itself and in this case of low-level cognitive activity cognition can be accounted for in terms of merely reflex-based notions: no other

[4] Traditional classical studies that see life as cognition and cognition as life are due to Maturana and Varela (1980; 1987; 1988; 1979; 1997) as also emphasized by Stewart (1996).

internal more or less stable representations are present. It is mandatory to note that the seminal studies that Schrödinger's performed on energy, matter, and thermodynamic imbalances furnished by the environment, address the attention to the fact that all organisms, also bacteria, are capable to perform simple *cognitive functions* because they "sense" the environment and process internal information for "thriving on latent information embedded in the complexity of their environment" (Ben Jacob et al. (2006, p. 496).

Indeed Schrödinger contended that life involves the consumption of negative entropy, i.e. the use of thermodynamic imbalances in the environment. As a member of a complex superorganism—the colony, a multi-cellular community—each bacterium is endowed with the ability to sense and communicate with the other units of the collective and executes its work within a distribution activity. Hence, bacterial communication implies collective sensing and cooperativity through interpretation of chemical messages, differentiation between internal and external information, and a kind of self vs. non-self distinction (peers and cheaters are both in action). In this view "biotic machines" are *meaning*-based kinds of intelligence to be contrasted with the *information*-based forms of artificial intelligence: biotic machines create new information, attributing contextual meaning to collected information. Self-organizing organisms like bacteria are afforded—through an actual cognitive act—by "relevant" information that they later on couple with the regulating, reconstituting, and *plastic* activity of the contextual information (intrinsic meaning) previously internally stored, which reverberates the intra-cellular state of the cells. Of course the "meaning production" that occurs in the above processes regards structural aspects of communication that cannot be compared to the particular sentential and model-based cognitive abilities of humans, primates, and other more rudimentary animals, but still partakes fundamental functions with these like sensing, information processing, and collective abductive contextual production of meaning.[5] As stressed by Ben Jacob, Shapira, and Tauber

> In short, bacteria continuously sense their milieu and store the relevant information and thus exhibit "cognition" by their ability to process information and responding accordingly. From those fundamental sensing faculties, bacterial information processing has evolved communication capabilities that allow the creation of cooperative structures among individuals to form super-organisms (Ben Jacob et al. 2006, p. 504).

I have just illustrated the role of "cognitive" activities in non-human animals: another—more controversial—step, from cognition to computation, can be taken. Biological systems in order to survive must have agency which makes their natural computation[6] way more complex than the one constituting behaviors of pebbles—they have memory (in their physical body and especially in their DNA), they communicate actively with other living organisms (bacteria "talk" using chemical language)

---

[5] I have provided further examples that illustrate the role played by cognition in animals and plants in chapter five of (Magnani 2009).

[6] On the concept of "natural computing" and its variegated meanings see below Sect. 4.6.

and exchange DNA (learn) through an information processing activity that refers to computation. Hence, also in this computational perspective, a difference between bacterial ignorance and the ignorance of a pebble should be established.

4.1.2 Ignorance in an Eco-Cognitive Perspective

I think that the main advantage of speaking of the transformation of "ignorant" entities in cognitive ones through computation derives from its capacity to stress the dynamic character of processes of attribution of cognitive roles to external entities. This idea is certainly related to the studies in the area of evolutionary theories—especially the ones on cognitive niche construction[7]—and to the recent tradition of studies in the area of cognitive science called "distributed cognition". Indeed, the idea of cognition that I am adopting in this chapter is informed by the "ecological" perspective I fully explained and described in my recent book *The Abductive Structure of Scientific Creativity. An Essay on the Ecology of Cognition* (2017). In this ecological perspective individual human agents, senses, environment, and cognition are all involved. As just mentioned we also need to refer to the cognitive science tradition of embodied and distributed cognitive systems: the emphasis in this case is on the "practical agent", on the individual agent operating "on the ground", that is, in the contexts of actual life.[8]

Moreover, cognitive processes are better understood if seen situated in and distributed across real socio-artifactual settings, beyond the internalist view typical of the traditional theories in cognitive science: cognitive activities are very often "delegated" to various (eventually artifactual) entities of the external environment and problem solving—for example—is in general occurring in collaborative contexts. The eco-cognitive perspective stresses the role played by the agent-environment interaction: for example, it is well-known that in current collective work environments, humans and artifactual technologies are intertwined in the activity of preserving and managing representational states, very frequently to the aim of solving problems.

Ideally, and as a first approximation, in my perspective a cognitive system can be seen as a triple (A, T, R), in which A is an *agent*, T is a *cognitive target* of the agent, and R refers to the *cognitive resources* on which the agent can rely when engaged in meeting the target—information, time, and computational capacity, to indicate the three most relevant. As I have sketched before my agents are also *embodied distributed cognitive systems*: cognition is evidently embodied and the interrelation between brains, bodies, and external environment are its fundamental features. In this perspective cognition is seen happening taking advantage of a continuous exchange of information in a complex distributed system that goes across the boundary between humans, artifacts, and the surrounding environment, where

[7] A further note about these theories is illustrated in Sect. 4.7.1.

[8] In chapter one of this book I have illustrated some fundamental concepts concerning distributed and embodied cognitive systems also describing the seminal ideas proposed by Hutchins (1995).

also instinctual and unconscious capacities play a crucial role. At this point I have to add that Peirce himself anticipated the approach to cognition I have just described by interpreting the concept of "inference" *semiotically* (and not simply "*logically*"), saying that all inference is a form of sign activity, where the word sign includes "feeling, image, conception, and other representation" (Peirce 1958, 5.283). It is clear that this semiotic perspective, which sees signs as distributed everywhere, is substantially consistent with my view of cognitive systems as embodied and distributed systems.[9]

Finally, I have to note that, as I plainly explained in chapter one, the concepts of information processing and computation are in our era certainly involved in the definition of cognition, but to elucidate what kinds of computation and information processing are implicated in cognition not only is a difficult task, but it is destined to become outdated because the concept of computation evolves and mutates, and the other two concepts, intertwined with it, and already marked by volatile and multiple definitions, subjected to the developments of knowledge, technologies, and cultural frameworks, also evolve. It is in this perspective that we can exploit the concept of ignorance: the transformation of ignorant entities in cognitive entities through computation is intertwined with the dynamic and emergent character of processes of attribution of cognitive—and computational—roles to external entities.

The reader does not have to misunderstand my claim: obviously, even if knowledge about cognition and its relation to information processing/computation is steadily increasing, that does not prevent us from looking into the present state of the art of the domain knowledge within research fields that study it. It is well-known that there are extremely quickly developing fields such as for example machine learning or AI, but the very fact that they are rapidly evolving does not prevent us from talking about them.[10]

4.2 An Eco-Cognitive Perspective on Computation

In chapter one I have introduced the idea of *Eco-Cognitive Computationalism*: computational activities have to be seen in context, taking advantage—as I explained in the previous subsection—of the recent contributions provided by cognitive sci-

[9] In subsection 4.4.2 below, I will exploit some Peircean concepts to explain how a physical computational system itself becomes a special domesticated actuator of semiosis.

[10] Neural information processing is the whole research area with dedicated conferences, such as NeurIPS 2019, Thirty-third Conference on Neural Information Processing Systems [https://neurips.cc/Conferences/2019] and ICONIP2019, The 26th International Conference on Neural Information Processing [http://ajiips.com.au/iconip2019]. ICONIP2019 aimed in a very intellectual wide way at presenting professional research results and enthusiastic discussions among researchers, scientists, and industry professionals who are working in neural network, deep learning, and related bio-inspired computing fields. The theme of the conference was Human Centred Computing: use of human sensor data to predict or model human emotion, intentions, and goals with an overall aim to produce computer systems which are useful, usable, and sympathetic to users. Apart from the field of neural information processing/neural computation there is a whole field of material computing, active matter, programmable matter or similar.

ence to the fields of embodied, situated, and distributed cognition. In this perspective computation has to be seen as occurring in "external" *physical entities* suitably metamorphosed so that they acquire the status of what I call *cognitive mediators*,[11] in which data can be encoded and decoded to reach interesting outcomes.

As already repeatedly illustrated, eco-cognitive computationalism, even if sees cognition as computational, does not reduce computation to digital computation. As I have illustrated in chapter one I privilege a perspective that sees the evolutionary emergence of information, meaning, and of the first vestigial forms of cognition, as the result of a complicated eco-cognitive interaction and simultaneous coevolution, in time, of the states of brain/mind, body, and external environment. This view does not contemplate a stability of the meaning of concepts such as computation, information processing, and cognition. The concept of computation I am assuming here goes beyond digital computation, that is the processing of strings of digits according to rules; it also concerns other computations executed by physical devices, such as for example brains, with their activity of modifications of the configurations of neural networks, but also analog computers, and the so-called neural computers. It is an expanded idea of the concept of computation: as I have indicated above the concept does not have to be seen as static, it develops according to the changes of theory and practice.

Let us repeat, this eco-cognitive view of computation is firmly intertwined with studies on distributed and embodied cognitive systems, which see cognition as a socially distributed process, oriented by actions and taking advantage of material artifactual entities, in which the "ecological" view also stresses the effects of the agent-environment interaction. The ecology of cognition framework makes us able to focus on "physical computation", exactly following Turing's original ideas concerning the emergence of computation in organic, inorganic, and artefactual agents. This perspective is especially appropriate to deal with morphological computation because its main tenet is that computation is "physical", as it is occurring in external entities.

Following my eco-cognitive perspective we can see "innocent"[12] (and "ignorant") entities, in the sense of pure tools or instruments "devoid of cognitive capacities", becoming first of all bearers of information and knowledge, and, finally, of computation, with the birth of both Turing's (Universal) Logical Computing Machines (LCMs) (which sees computation as a merely syntactic process) and the strictly intertwined concrete (Universal) Practical Computing Machines (PCMs). To make an example, blackboards are simple non technological artefactual entities, which can be endowed with cognitive delegations thanks for instance to the drawing of diagrams or the writing of texts. More technologically complex artifacts, such as telephones and telegraphs, were able to carry and perform elementary cognitive tasks, but it is only with highly technological man-made instruments, such as digital computers,

[11] I have proposed this expression in (Magnani 2001), for more details cf. footnote 18, chapter three.

[12] The reasons that explain the adoption of this adjective, endowed with a moral halo, are illustrated below in this section.

that we face the final "loss of innocence" of entities. Turing would have said that these technological tools become *universal* cognitive physical entities, in so far as they become *computational* artifacts that compute for humans or artefactual agents.

When I say loss of "innocence" I am not equating innocence to mere cognitive passivity. I want to stress the fact that when ignorant entities are transformed in cognitive and, at the same time, in computational ones, they become endowed with a moral worth and can play a moral role (and so moral acts and possible consequent violent acts). As I have illustrated in my books concerning philosophy of technology (Magnani 2007) and philosophy of violence (Magnani 2011), modern technology—computational artifacts obviously included—has brought about (potential) moral and (potential) violent consequences of great magnitude. Imagine the consequences (often unintended) of technological tools such as internet and artificial intelligence, for example the negative effects on privacy in cyberspace and the threats to individual identities, cultures, etc. Computational structures and technological artifacts became moral carriers and mediators of effects that can be perceived and/or judged as violent, strictly integrated with humans, and so surely less "innocent" than simply ignorant entities.

Finally, I have to again repeat that eco-cognitive computationalism does not aim at furnishing an ultimate and static definition of the concepts of information, cognition, and computation, instead it intends, by respecting their historical and dynamical character, to propose an intellectual framework that depicts how we can understand their forms of "emergence" and the modification of their meanings: as I have already said in the previous subsection it is in this perspective that we can see the transformation of ignorant entities in cognitive entities through computation by stressing both the dynamic and emergent character of processes of attribution of cognitive—and computational—roles to external entities.

4.3 Computation Is Physical: The Epistemological Overturning

It is well known, thanks to the cognitive studies of the last decades, that a remarkable simplification of cognitive and motor tasks in human and animal beings is often performed thanks to their morphological aspects—and this character had been already obviously used in robotics. The recent novelty is that we can exploit morphologies of appropriate entities to integrate classical computation: in this way we construct what I called *mimetic bodies* (see above, chapter two, this book) able to simplify the whole process of computation of a certain task, so fulfilling a general request of that *simplexity*[13] which characterizes animal embodied cognition, that is the complementarity of complexity and simplicity (Kluger 2008; Berthoz and Petit 2014). To better

[13] Simplexity regards the possible complementary relationship between complexity and simplicity in the framework of a dynamic interplay between means and ends (cf. (Berthoz and Petit 2014)).

understand this novelty—and other new unconventional computational cases—it is necessary to establish a naturalistic view of computation, acknowledging the fact that it happens through physical systems, technologically appropriately built and tested.

4.3.1 Computational Domestication of Ignorant Physical Systems

In chapter one of this book, I afforded—taking advantage of the intertwining between the conceptual interplay between information, cognition, and computation—the following classical problem: "Is cognition computation?" We can answer: yes it is, but cognition is not exclusively computation. A variety of serious studies concerning the many ways in which information processing and computation are intertwined with cognition has been recently produced, even if in the perspective of the ecology of cognition I previously sketched they become outdated. I said (and illustrated in the first two chapters) that the notion of computation changes from the point of view of the meaning and the same happens in the case of the other two concepts. All three concepts can be usefully captured by definitions, that anyway are subjected to obsolescence due to the change of the developments of knowledge, technologies, and cultural frameworks. Several studies have elucidated various notions of computation in the light of recent epistemological and cognitive science research (see for example (Vallverdú i Segura 2009)) and emphasized that computation has in general to be seen as enforced in *physical systems* (of course avoiding to arriving to the so-called pancomputationalism,[14] that to a first approximation consists in the view that every physical system is a digital computing one and can be illustrated in computational terms).

Vallverdú is also inclined to attribute a relevant epistemological dignity to computational modeling (he calls this aspect "computational epistemology"), not only because scientific knowledge (for example creative reasoning) is currently obviously performed/mediated by computational tools and models, but also for its capacity to "directly" produce knowledge. He even says: " 'Knowing' no longer constitutes the strict scope of the human mind (Clark 2003). The automatization of the problem of the four colors or of the conjecture of Kepler, both already theorems, are examples of this new form of obtaining knowledge" (Vallverdú i Segura 2009, pp. 565–566). In a similar and even more optimistic vein (Kari and Rozenberg 2008, p. 83) observe:

> In these times brimming with excitement, our task is nothing less than to discover a new, broader, notion of computation, and to understand the world around us in terms of information processing. Let us step up to this challenge. Let us befriend our fellow the biologist, our fellow the chemist, our fellow the physicist, and let us together explore this new world. Let us, as computers in the future will, embrace uncertainty. Let us dare to ask afresh: "What is computation?", "What is complexity?", "What are the axioms that define life?" Let us relax our hardened ways of thinking and, with deference to our scientific forebears, let us begin anew.

---

[14] Cfr. also below, Sect. 4.7.

This perspective on physical computation has to be further deepened with the adoption of the eco-cognitive framework, that depicts how we can understand the "emergence" of what I call the *new* computational "domestication" of physical entities, thanks to morphological computing. I have illustrated in detail the various aspects of physical computation in chapter one above (section 2.3.1) by indicating various insights on the problem of physical computation as illustrated in (Horsman et al. 2014, p. 14). Here I list again the various items, for the ease of the reader:

1. a *computer* is a physical system composed by concrete parts and by its peculiar inner connections that allow passages from one physical state to another;
2. Horsman et al. contend, and I agree with them, that physical computing is "the use of a physical system to predict the outcome of an abstract evolution" (Horsman et al. 2014, p. 14);
3. the physical entities devoted to computation are *technologies*, fruit of scientific knowledge: step after step they became engineered physical systems which are endowed with "wanted" features;
4. the physical computation is successfully concretized when we are capable to enforce a precise relationship between abstract mathematical/logical entities and the physical ones;
5. of course the notion of *computation*, and its system property, *information*, can be used in other intellectual fields of research, trying to "explain" processes such as for example photosynthesis or the conscious mind, and some researchers arrived to contend that "everything is information" or "the universe is a quantum computer" or "everything computes" (Horsman et al. 2014, p. 2). In this case I fear that the notion of *physical computation* tends to become devoid of meaning: not all physical systems (biological and artefactual included) can carry computation. I will come back to this issue below in Sect. 4.7.

4.3.2 The Epistemological Overturning: Physical Computing Versus Physics

I have already illustrated in more detail, in chapter two of this book, that *a physical system is used to predict the outcome of an abstract dynamics*: this is the opposite of what happens in physics, in which *an abstract model is used to predict physical dynamics*. Horsman et al. distinctly contend (2014, p. 2) that physics works by representing physical systems abstractly, using abstract theories to predict the result of physical evolution, and computation uses *a physical system to predict the outcome of an abstract dynamics*. The ignorant technological entity (for example the digital computer) has been domesticated to become a tool that is endowed with computational—and so cognitive—capacities: it works because the *predictions* concerning its behavior are reliable, exactly as in the case of an equation that is a good explanation of a physical system and provides good predictive abilities.

Of course computers are not always used to predict. For example, the operating system is used to control processes in the computer. Computers today are used a lot to generate behavior. Some are used to operate databases. Some computational processes are run to show movies, play music, communicate and enable communication. Some are simply calculating values of given equations for given inputs. While performing calculation, computer is typically not predicting. So if a process fails to predict it still can be a computational process. Wolfram and Chaitin talk about a computing universe that computes its own next state. It does not predict as it is an extremely complex process, but it unfolds in real-time. To avoid misunderstandings[15] I have to make clearer that Horsman et al.'s idea of a computer as a *physical system used to predict the outcome of an abstract dynamics* does not refer to the final performance or "use" of a computational program, that not necessarily regards predictive performances. It just refers to the fact that what I call an ignorant technological entity can be domesticated to perform a "physical behavior" which it is itself to be considered *predictable* and so reliable.

Obviously, as already remarked in chapter two of this book, a *representational process* is also necessary, able to assure a "comparison between physical processes and mathematical described computations" (Horsman et al. 2014, p. 2); again, in computation "the initial impetus is not a physical system that needs to be described, but rather an abstract object that we wish to evolve. An abstract problem is the reason why a physical computer is used" (cit., p. 10).

4.4 Enhancing Ignorant Bodies Through Disseminated Computation: The Birth of *Mimetic Bodies*

4.4.1 Classical Computational Physical Systems Are Fruit of Technological Domestication

To render an ignorant entity such a sheet of paper cognitively significant is relatively simple, it is sufficient to write some statements using a natural language. The tradition of cognitive science concerning the so-called "distributed cognition" would add that in this non-computational case we are delegating to an external object a cognitive role, as an anchor for our mind (Mithen 1999; Fauconnier and Turner 2003; Hutchins 2005) or as a tool for rendering our mind "extended" (Clark and Chalmers 1998).

In the case of computation, Turing himself clearly said that a simple "Paper Machine" is able to compute: "It is possible to produce the effect of a computing machine by writing down a set of rules of procedure and asking a man to carry them out. [...] A man provided with paper, pencil and rubber, and subject to strict discipline,

[15] This paragraph has been motivated by an objection to a passage concerning computation and prediction of a previous article of mine by an anonymous reviewer.

is in effect a universal machine" (Turing 1969, p. 9). However, we know that for more efficient computational results a more complicate ignorant but potentially cognitive entities have to be built: they are the fruit of a technological construction.

I have always considered it incredible that researchers have been able to embed the Turing's abstract machine, as a Discrete-State Machine (DSM) in physical machines, so building artefactual technological engines that are capable to evolve by discrete states, thanks to the virtues of the silicon electronic structures.[16] It is well-known that it was science that provided those conceptual tools to engineer physical systems that humans later on "domesticated" to get technologies characterized by "wanted" features and expected behaviors. In the case of digital computers researchers started from a good physical theory, capable to delineate an abstract description of the physical system to be built by technology. That new built artefactual system had to perform a set of evolutions in expected ways, to be computationally reliable.

It is important to stress that non-standard physical artefactual entities that are exploited as computing tools and substrates—with the exclusion of quantum computers—present a less mature theory so that their reliability is lower than the one of silicon devices (Horsman et al., 2014). However, I have to add that that which Horsman et al. see as a problem, Kari and Rozenberg, in "The many facets of natural computing" (Kari and Rozenberg, 2008) instead see as inspiration and challenge of two-way learning processes endowed with a heuristic worth. To make an example, we learn about neurons applying information-processing model on them and comparing to the empirical results. We adjust both the model of the physical system and the model of computation, recursively, and improve both in the process.

We will soon illustrate that the new non silicon-based substrates that take advantage of the resources offered by their morphology are becoming computationally significant. As for now, more words about the problem of encoding and decoding have to be added, a process we can call of *semiotic domestication* of the already technological domesticated artefactual physical systems.

4.4.2 Semiotic Domestication: *Encoding, Decoding, Computational Entities, Unconventional Substrates*

It is the only presence of a "computational agent" that permits the exploitation of the already domesticated (thanks to science and engineering) physical computational entity (but still ignorant) as a machine that performs cognitive tasks in the external environment (see also Sect. 2.3.4 of chapter two, this book).

1. First of all the agent starts a computational process by *encoding*—that is by *representing*—abstract data in the artefactual physical system (as I have said, we already have a physical theory that grants that the artefactual physical system

[16] The reader that is interested in the interplay Discrete State Machine (endowed with a Laplacian behavior) and continuous systems in computer science can refer to (Longo 2009a, b) and to my recent (Magnani 2018a, b).

appropriately operates: we expect the encoded data will work). Encoding is a way of managing a problem, thanks to some appropriate cognitive representation, for example mathematical, to have it encoded and at the same time channeled by abstract algorithms, in turn encoded (they become concrete algorithms) into the machine: at this point the artefactual physical entity starts to behave as a computer. It is through this process of further domestication that we can reliably expect to get the computable output of abstract evolutions, obtaining what Turing called *(Universal) Practical Computing Machine* (PCM).

2. Charles Sanders Peirce clearly illustrated that humans create signs, a word for example, or an icon, which refer to some external object or event: this is called "process of semiosis". Still according to Peirce, we can say that human brains build a lot of signs that are exploited to the aim of responding to a series of other signs and to expect some feedback. It is being involved in this process, which is material and semiotic but also clearly cognitive, that human brains become "minds" to produce intelligent reasoning. We perfectly know that signs can be externalized in both natural and artefactual mediators, for instance thanks to drawing and writing, originating that process of externalization of the mind (or disembodiment of the mind), which has been studied by cognitive science as one of the main aspects of the idea of distributed cognition. It is in the interplay between internal and external representations that a considerable part of standard human cognition is occurring, but also that part of cognition that consists of more creative results, such as the generation of new thoughts, ideas, and concepts that do not have a "natural home" internally. In this perspective minds have to be considered as "extended" and artificial in themselves, given the role played by external representations.
3. The activity of encoding described in the first item is still a process of semiotic externalization of signs: it is the *incorporation* of a problem (semiotically expressed thanks to words, but eventually also symbols, technical languages, icons and other representations), with the help of the mediation of some appropriate other cognitive tools already embedded in the engineered physical artifact (basic programs, algorithms, etc.) In this way the system itself becomes a domesticated actuator of semiosis, computationalized, in this case. Furthermore, decoding is the process of picking up the meaningful signs produced by the "concrete evolution" of the physical artefactual system.
4. When thanks to the physical process of our doubly domesticated artefactual physical system we arrive at the final physical state (rendered possible by the physical evolution of the system, usually a classical silicon device), a final state is *decoded* by the computational agent (a human or another machine) as a new abstract semiotic state. We have to remember that to get this outcome we have to face with a concrete evolution, not an abstract one. Without the encode and decode aspects, there is no computation, there is a trivial "ignorant"—so to speak—physical system undergoing evolution: "going about its business", *only potentially computational* (Horsman et al. 2014, p. 5). In this light computation can be clearly illustrated as the fruit of the complex composed system emerging from the processes of both technological and semiotic domestication.

Interestingly, as already stressed in chapter two above, "This is how we can escape from falling into the trap of 'everything is information' or 'the universe is a computer': a system may potentially be a computer, but without an encode and a decode step it is just a physical system" (Horsman et al. 2014, p. 15). I have to clearly note that the process of encoding and decoding refers here to an act of delegation intentionally performed. As recently remarked by (Horsman et al., 2017a, b) this is also happening in the case of the "computational domestication" of unconventional substrates such as in the case of DNA computing. Indeed, how about DNA computing going on in the cell? This does not have to be considered as an "objective" natural process of encoding and decoding, but just another case of computational delegation to biological entities that render possible that kind of computational/cognitive domestication I am illustrating in this chapter. I will provide further details on this issue below in the Sect. 4.6.1 "Information Units Unconventionally Computed by Biological Systems".

5. Hence, coding and decoding acquire meaning and consistency only if seen as related to a *computational entity*, human or artificial, that works as a semiotic delegator and recognizer of *meanings*, when dealing with the specific artefactual physical entity at play.

4.5 Morphology-Based Enhancing: Computational Domestication of New Substrates

Organic living beings of all kinds can be considered physical entities that are potentially inclined to process information (more in general, they are "semiotic beings", Peirce would have said) and that can "also" potentially be exploited to perform computations. The example of Turing's paper machine, which regards the chance human living beings have to become computational beings – I have quoted above in Sect. 4.4.1—is mandatory.

At this point it is important to stress that entities that are not comparable to the engineered silicon devices, but that have the morphological features typical of the bodies of organic beings, offer a "new" way of becoming part of a computational process. I will illustrate below that research on physical RC—Reservoir Computing—represents a new way of using the dynamical properties of body parts to channel important computational tasks.

Of course a great relevant and reliable quantity of scientific and technological knowledge about the traditional engineered silicon substrates is available. Therefore, if we desire to have at our disposal new substrates able to carry computation it is necessary to provide adequate physical theories of them, because we have to be assured they will present regular behaviors: we need theories able to predict expected modifications of the substrate that are in turn empirical confirmed. A stone, a soap bubble, a large interacting condensed-matter system, or other entities can execute some physical processes and the, after a while, the termination of the process can be established and some measures extracted, but no guaranties we have that we face a computational process. In summary, it is not good to establish the computational description of a physical evolution post-hoc. We always have to look for a more reliable computational domestication of new ignorant substrates: morphological computing is one of the recent answers.

4.5.1 Morphology-Based Enhancement of Mimetic Bodies

In Sect. 2.3.5 of chapter two how morphology can be exploited at the computational level. For the ease of the reader we list again the main way of exploiting morphology:

1. morphology that helps control, in which no control system, no motors and no sensors are at play;
2. morphology that helps perception;
3. morphological computation in proper sense, for example "reservoir" computing (RC), in which embodiment and computation are rigorously intertwined.

The passive dynamic walker represents an example of the first case (the behavior is happening exclusively by mechanical interaction); the gecko feet are an example of the second case, which can also be considered as active extensions of the passive walker above, thanks to actuators and sensors (their special ability is the consequence of their morphology interacting with specific surroundings—and so not principally that of a higher-level central control); the case of the eye of the fly too, which not only regards movement, but also cognitive capabilities related to perception, is another typical good example; it is Reservoir Computing that incarnates the third most interesting case, theoretically studied and advanced by the neural network community, and amazingly applied to physical entities/devices that can work as a reservoir ("Physical" RC), in which the body is efficaciously domesticated for computation.[17] Various physical entities can work as a reservoir, giving birth to the so-called Physical RC, which can also be the body of an agent: biological bodies which act in the environment can be endowed with the required properties—high dimensionality, nonlinearity, and fading memory, which render them suitable for playing a computational role.[18]

4.5.2 The Birth of Mimetic Bodies: Enhancing Ignorant Bodies Through Disseminated Computation

To further illustrate the modes of domesticating—beyond the well-known case of silicon devices I have described above—esoteric entities for computation, it is important

[17] Further details are illustrated in Sect. 2.3.5 of chapter two, this book.

[18] More details are provided by (Müller and Hoffmann 2017, p. 3) and (Hauser et al. 2011, 2014).

to clarify the main cognitive and epistemological features of RC (Reservoir Computing). To make an example, imagine you have at your disposal a traditional "digital" computational controller tool (not necessarily a PC, but also a robot)—that already presents its own "traditional" engineered physical embodiment—you want to simplify and improve from the point of view of its capacities and/or behavior. You can do this thanks to some cognitive "virtues" that are outsourced to the morphology of physical entities with the aim of receiving an estimable help in terms of what can be called "analog" computation. Physical RC regards real physical bodies that are exploited as reservoirs and as computational generators. In the case of both entities, the extremely engineered silicon device and the Physical RC, they do not "know" that they belong to a computational device. They trivially conform to the laws of physics and they directly answer, once computationally domesticated. In the case of Physical RC, it is important to add "[...] that the body does not over—nor under—compensate, since it is a stable physical system. The proposed setup simply adds some linear readouts to the body to complete the computation. The body would react exactly the same, if there were no readouts at all. If this output is used, e.g., in a feedback loop as a control signal for the robot, the behavior of the robot of course should be different if we close the loop by adding the readout" (Hauser et al. 2014, p. 230). Moreover, in morphological computation the physical computer does not need to be intelligently conceived and engineered: it naturally evolves and this evolution also provides a computation.[19]

As already observed in chapter two of this book, the body is "ignorant" (Hauser et al. 2014, p. 231), it behaves the same whether we add readouts or not and, this is remarkable, can be exploited in several computations on the same input at the same time supplementing the suitable readout. Classifying the body as "ignorant" of the computations that are channeled through it may be strange. It is more than obvious that no part of the entity acknowledges that is part of the computational system itself, it simply presents a kind of passivity to the computation that is occurring through it. It is ignorant of the semiotic life it is computationally channeling but at the same time it has been domesticated to do it so appearing to the *computational agent* (the user, see above Sect. 4.4.2) a cognitive smart entity.

The traditional "physical" computational entities (hardware) as silicon devices and robots empowered with an abstract controller are founded on the discreteness of *digital* processing and they appear, at least phenomenologically, from the rough exterior perspective, as stable, insusceptible, and robust with respect to the external influences that are not ascribable to encoding and decoding (see above Sect. 4.4.2). Morphological computation, in the case of Physical RC, is instead happening in the continuous actual reality (without digital complicatedness) as a limpid kind of quick and simple *analog* computation, in which the use of concrete entities at the same time obviously causes restrictions and limits. It is certainly characterized by an absence of a great cognitive/computational plasticity (and consequently of "universality", if compared to classical Turing machines).

[19] On this issue, cf. also (Horsman et al. 2014, pp. 19–22).

Thanks to morphological computation the bodies that are "ignorant" become *mimetic bodies*, that is, bodies that are able to mimic several cognitive tasks because embedded in a computational environment. The term mimetic that I am using here is in tune with some of Turing's ideas: indeed he says that by "mimicking education, we should hope to modify the machine until it could be relied on to produce definite reactions to certain commands" (Turing 1969, p. 14). Analogously, mimicking the morphological aspects of an entity we obtain computational outcomes: a notable case is the already quoted one illustrated by Nakajima et al. (2013, 2015), who have "rigorously" mimicked part of the morphology of an octopus exploiting a prototype of a soft robotic arm suggested by it.

In the case of morphological computation we are dealing with new relevant epistemological and technological advances of what I can indicate as *disseminated computation*: we are facing a computation increasingly distributed in an extended diversity of props, tools, bodies, and devices to the aim of getting new cognitive results. A further application could be imagined in the case of robot design to the aim of constructing an unexpected generation of robots with increased adaptability and a minor amount of involved control parameters.

4.5.3 Mimicking Morphology and Synthetic Biology

A final remark on the concept of mimetic body has to be added. Understandably, mimicry, at a general cognitive level, can be considered a morphological wired mechanism that makes possible learning, and embeds intentionality. In the present chapter I have illustrated a more restricted sense of the word: thanks to morphological computation bodies that are "ignorant" become *mimetic bodies*, that is, bodies that are able to mimic various cognitive tasks because inserted in a computational context. I added that the term mimetic is in this case coherent with Turing's conception: he says that "to modify the machine until it could be relied on to produce definite reactions to certain commands" (Turing 1969, p. 14) it is necessary to "mimic" education, a metaphor that introduced a new idea surely at the roots of the modern concept of computational program. In sum, in our present case morphological computation efficiently mimes some cognitive processes thanks to the integration of the morphology of entities in a traditional computational digital context.

Finally, it is important to remember again that, in 1952, though, Turing too dealt with the problems of morphology (Turing 1952). He indeed published an article on morphogenesis which presented a very original, but non computational, non-linear system of action-reaction and dynamic diffusion, in which he proposed what he called a model of the physical phenomenon in question. He thus sought to propose a structure of determination, by means of the equations describing causal interaction in the action-reaction process (Longo 2009a).

At this point it is mandatory to ask a question regarding the role of morphologies, which affords a specular problem with respect to the one of morphological computation: what happens when, on the contrary, we need to mimic some (biological) morphological entities? Synthetic biology inaugurated an ambitious program aiming at answering this question. With synthetic biology we are dealing with "alternative realities", synthetic fictions, we can say, in which, as contended by some epistemologists, synthetic "fictional" imagined systems are turned into "concrete" ones (Knuuttila and Loettgers, 2017). At least three main areas characterize this research: (1) the project of constructing a minimal genome (that would include the smallest number of genetic elements sufficient to produce and sustain a free-living cellular organism in an ideal environment); (2) the building of an artificial and minimal cell as "a hypothetical biological system that possesses only the necessary and sufficient attributes to be considered alive" (Gil 2011, pp. 1065–1066), also aiming at the construction of protocells or the so-called chassis cells;[20] and (3) the synthetical design of alternative genetic systems as chemical structures.

Given the fact that many difficulties are experienced by synthetic biologists when they apply plan engineering models to biological systems and consequently real biomachines fail because of the lack of operativity at multilevel complexity,[21] computation can be still of help, but differently than in the case of morphological computing. However, in these kinds of research regarding the systematic design and engineering of biological systems the problem of managing engineering complexity can be afforded thanks to the involvement of artificial intelligence. Biologists have started to institute engineering control over the genetic machinery of cells and identified and generated many DNA sequences which can be exploited as building blocks of new cellular programs. These "programs" are then injected into cells and enforced. The studies have arrived at a complexity threshold that AI researchers can help overcome.

The current procedures in synthetic biology demand practitioners to project organisms at the DNA level. These processes are mainly manual and become too complex when the dimension of the project grows. These kinds of processes can be considered analogous to writing a computer program in assembly language which itself turns difficult as the size of the program grows. As Yaman and Alder (2013) further say: "Currently the synthetic biology engineering workflow is mostly manual and relies heavily on domain expertise, a limited amount of which is shared through publications. There are several points in the workflow where informed decision making would improve the efficiency and reduce the time to engineer an artificial genetic circuit. Such decision points are where the biologist can use the help of AI researchers". AI researchers can help synthetic biologists with reasoning, multiagent systems, and robotics taking advantage of results in the areas of machine learning, knowledge-based systems, knowledge representation and reasoning, reasoning under uncertainty, heuristic search and optimization. To conclude, even if biological organisms are complex and not completely understood, there are various opportunities for AI techniques

[20] On JVCI-syn3.0 as "a working approximation of a minimal cell" see (Hutchison et al. 2016, p. aad6253-1).

[21] Indeed synthetic organisms are currently designed at the DNA level, which limits the complexity of the systems.

to make a major difference in the efficacy of organism engineering.[22] In this last case we deal with the reverse of the role of computation I have described in this chapter and in chapter two above when illustrating morphological computing: indeed it is a computation (that we can call "natural computation") that aims at miming (biological) morphological entities and not a morphology that supplies computation in a digital environment that can take advantage of it.

Finally, coming back to the problems of computationally domesticating new substrates, it is interesting to refer to the following the recent development of biocomputing, described by Grozinger et al. (2019). The so-called "cellular supremacy" refers to a set of problems that biocomputers can practically solve but that traditional microprocessor-based devices cannot, there is a kind of "added value" that living systems offer beyond silicon-based hardware:

> Many current implementations of cellular computing are based on the "genetic circuit" metaphor, an approximation of the operation of silicon-based computers. Although this conceptual mapping has been relatively successful, we argue that it fundamentally limits the types of computation that may be engineered inside the cell, and fails to exploit the rich and diverse functionality available in natural living systems. We propose the notion of "cellular supremacy" to focus attention on domains in which biocomputing might offer superior performance over traditional computers. We consider potential pathways toward cellular supremacy, and suggest application areas in which it may be found.[23]

4.6 Natural Computing: Computing Inspired by Nature and Computing Occurring in Nature

Morphological computing is just one example of the exploitation of new substrates for computation: a recent handbook (Rozenberg et al. 2012) on the so-called *Natural Computing* illustrates both human-designed computing inspired by nature and computing (and information processing) occurring in nature. Examples of the first kind of research described in the handbook include neural computation inspired by the functioning of the brain; evolutionary computation inspired by Darwinian theory of evolution; cellular automata inspired by intercellular communication; swarm intelligence inspired by the behavior of groups of organisms; artificial immune systems inspired by the natural immune system; artificial life systems inspired by the properties of natural life in general; membrane computing inspired by the compartmentalized ways in which cells process information; and amorphous computing inspired by morphogenesis.

Among the studies which aim at substituting traditional silicon devices with new substrates, which I am illustrating in this chapter taking advantage of the case of

[22] A 2018 AAAI Fall Symposium has been devoted to these studies: "Artificial Intelligence for Synthetic Biology", October 18–20, 2018, Arlington, Virginia, USA. See also (Yaman et al., 2018).

[23] Furthermore, on how dynamical systems in general "beyond the digital hegemony" store and process information as a fundamental question that touches a remarkably wide set of contemporary issues, cf. (Crutchfield et al ., 2010).

morphological computing, the handbook lists molecular computing and quantum computing. In molecular computing, information is encoded as biomolecules and then molecular biology entities are used to process the data, thus generating computations. In quantum computing quantum-mechanical phenomena are used to concretize computations and produce communications more efficiently than classical physics and, hence, in the case of standard hardware. Of course morphological computing represents an important aspect of this area of research, because it involves entities that become part of computation thanks to the features of their structure.

The second kind of research, regarding computation taking place in nature refers for example to the study of the computational nature of self-assembly, which lies at the core of nanoscience, the computational nature of developmental processes, of biochemical reactions, of bacterial communication, of brain processes, and the systems biology approach to bionetworks, where cellular processes are studied taking advantage of communication and interaction, and, hence, of computation.

We can finally observe that the fear by some critics of natural computing, of loss of generality (universality) of computing is not well grounded. We can have different kinds of computers, some of them universal, some of them specialized for specific computations. Quantum computers will be such specialized computers, far better than our present ones for certain applications, not as good for others. In this perspective computers are not that different from other tools and instruments. We would not think of constructing an apparatus that would at the same time be good as a microscope and a telescope. In the same way, we will also have computers that will be exquisite simulators of quantum systems but not useful for playing computer games.

4.6.1 Information Units Unconventionally Computed by Biological Systems

An interesting case of exploitation of biological entities to perform computation has to be stressed, taking advantage of my eco-cognitive metaphor. Special types of informational units can be computed by biological entities, units that can be found into unexpected and unconventional computing systems, different from the ones I have illustrated in the previous section concerning morphological computation. It is the case of the DNA computers, which represent another prominent kind of unconventional non silicon-based computation. The so-called molecular computing is a relatively recent area or research situated at the crossroad of various disciplines that created the tools and the methods to program molecules with the aim of getting a wanted computation, or build a desired object, or control the functioning of a given molecular system (Rozenberg et al. 2012, pp. viii–ix): "The central idea behind molecular computing is that data can be encoded as (bio)molecules, e.g., DNA strands, and tools of molecular science can be used to transform these data. In a nutshell, a molecular program is just a collection of molecules which, when placed in a suitable substrate, will perform a specific function (execute the program that this collection represents)" (*ibid.*)[24]

In sum, DNA computing is founded on the idea that molecular biology processes can be exploited to perform arithmetic and logical operations on information encoded as DNA strands (Hagiya et al. 2016). DNA bio-operations can be used for computations so that a biocomputation will consist of a succession of bio-operations: the DNA-encoded information is very different from silicon-based electronically-encoded information, and bio-operations obviously differ from electronic computer operations. Further details concerning this difference can be usefully found in Kari et al. (2012). On the very recent exploitation of various artificial molecular devices, including some made of DNA or RNA to develop robotic systems cf. (Hagiya et al., 2016). A molecular robot is composed by sensors, computers, and actuators, all made of molecular devices, and "reacts autonomously to its environment by observing the environment, making decisions with its computers, and performing actions upon the environment. Molecular computers should thus be the intelligent controllers of such molecular robots. Such controllers can naturally be regarded as hybrid systems because the environment, the robot, and the controller are all state transition systems having discrete and continuous states and transitions" (cit., p. 4).

To build what she calls our "DNA laptops" Kari (2013) usefully observes that the most important analogy that had to be discovered to open the door of DNA computing was the one between two processes, one biological and the other mathematical, that is: (a) the complex structure of an organic being is the consequence of the application of simple operations (copying, splicing, etc.) to initial information encoded in a DNA sequence, and (b) the result $f(w)$ of the application of a computable function to an argument w can be found thanks to the application of a combination of basic simple functions to w.

The success of the computational domestication of new substrates, as it is illustrated by the case of morphological computation, DNA and quantum computing, etc., also suggests some philosophico/epistemological considerations aimed to express some doubts about the consistency of positions such a pancognitivism, paninformationalism, pancomputationalism, relatively popular in the current debate concerning philosophy of computing. In the following final section I will describe those positions together with the concept of overcomputationalization in its relationships with the process of domestication of ignorant entities I have treated in this chapter.

---

[24] We also have to remember that important contributions of molecular computing also favored the understanding of some central issues of the nanosciences, for example self-assembly.

4.7 Protecting Ignorance: Beyond Overcomputationalization

4.7.1 Pancognitivism, Paninformationalism, Pancomputationalism

What I call *overcomputationalization* refers to the presence of too many entities and artifacts that carry computational tasks and powers, for example digital machines embedded in laptops, cell phones, servers, videos, cameras, social networks, robots, and also various unconventional substrates (I have described in detail in the previous sections the case of morphological computation), etc. Here the discourse becomes moral, and solutions are multiple. Whereas many would agree that the diffusion of computation is bringing havoc, some would say (like me) that it is caused by an excess of computation, others would content the opposite, i.e. computation has not spread enough.

After having introduced in the previous sections how our era is characterized by a progressive transformation of ignorant entities not only in cognitive entities but also in computational ones I would like to suggest some humble considerations concerning the potential excesses of this trend, that I think have to be analyzed and criticized. Indeed, I think that currently we are facing an excess of computationalization of the world and I am consequently complaining that too many ignorant entities are more and more "domesticated" (in the sense I have explained in this chapter). I will say more words on this issue, related to desire and suggestion of protecting the ignorance of some entities, in the last subsection of this chapter.

In the last decades philosophers have celebrated the computational revolution and positively evaluated many of its aspects and consequences. An interesting position that has been elaborated is pancomputationalism that from a philosophical perspective, as we will soon see, is in some sense the rightful heir of paninformationalism and pancognitivism. To better grasp my concerns regarding overcomputationalization it is useful to analyze the intertwining between these positions.

Pancomputationalist philosophers have imagined that everything is computing. The problem of pancomputationalism can be nicely synthesized thanks to the following argumentation, provided by Horsman et al. (2014, p. 2). To answer the question *when does a physical system compute*? is very difficult and no accepted answer is currently available. There is not a formalism able to tell us whether a computation is occurring physically, and various confusions affect the discussion concerning non-standard forms of computation. It is obviously accepted that a laptop processing a Matlab calculation and a server running a LaTeX text are physical systems operating a computation, but "However, when we move beyond standard and mass produced technology, the question becomes more difficult to answer. Is a protein performing a compaction computation as it folds? Does a photon (quantum) compute the shortest path through a leaf in photosynthesis? Is the human mind a computer? A dog catching a stick? A stone sitting on the floor?" If we say that all are computers we affirm that everything that possesses a physical status is executing a computation by virtue of its own existence. It is obvious to derive that from a conviction like this one, that sees a universe in which everything is a computer, the concept itself of physical computation becomes empty. Indeed "To state that every physical process is a computation is simply to redefine what is meant by a 'physical process' – there is, then, no non-trivial content to the assertion. A statement such as 'everything is computation' is either false, or it is trivial; either way, it is not useful in determining properties of physical systems in practice" (*ibid.*)

Pancomputationalism is often defended explaining that, when we state that everything is computational, we are simply using the word computational as a way of "interpreting" the system at stake, and of course everything can be interpreted using that metaphor. We have to say that this position creates a confusion between computational modeling and computational explanation, as I will soon specify in the following paragraphs. Other kinds of pancomputationalism, appropriately linked to paninformationalism, contend that the entire universe can be seen as computational and also arrive to maintain that information and/or computation possess a kind of priority with respect to physical materiality. Gordana Dodig Crnkovic (2011; 2013; 2017) proposes a richer info-computational and more circumscribed and integrated view as a synthesis of pancomputationalism—naturalist computationalism—with informational structural realism, and defends it by noting the fundamental role of computing in nature (natural computing).

As I have illustrated in chapter one, adopting an evolutionary view and exploiting some of Thom's ideas expressed in the so-called catastrophe theory (Thom 1988), paninformationalism is puzzling when seen in a naturalistic perspective. Let us reconsider the example already exposed in chapter two. When we are analyzing the case of an infection as a pregnance (carried by a virus, that is a material/biological mediator) that involves healthy agents, who represent the invested saliences that in turn can re-project that same contagion as a pregnance in a specific natural environment (where, in turn, other mediators such as air or blood are the transmitters), it seems bizarre to affirm that information (or even computation) is at play: biomedical scientific explanations of the phenomenon are already patently available and it seems superfluous to resort to other concepts such as the ones of information or of computation. One thing is to build a "model" of that event from an informational point of view or by using a computational program, and another is to produce a biological scientific knowledge of it. Of course this critique does not jeopardize the project (see for example the info-computational view, Dodig-Crnkovic (2017)) of enabling a view of nature also referred to computational modeling (information processing) thanks to a conjugation of all those specialized approaches in a perspective that does not aim at replacing natural sciences with computational models, but that aims at establishing a fruitful epistemological collaboration.

In chapter one above, taking advantage of the studies provided by Longo (2009a; 2012; 2017 I have described in detail the dangers that can arise by thinking that mimetic computational digital modeling "is" directly, ipso facto, scientific knowledge: for example, in the case of biological organisms, the gap between simulation and intelligibility is very strong, also because the variability is dominant.

A different view is advanced by Denning (2007): information processes and computation continue to be found abundantly in the deep structures of many fields. Computing is not—in fact, never was—a science only of the artificial (cf. also Denning and Martell (2015)). A good example of the interplay between natural science and computational modeling is given by Fisher and Henzinger (2007), who call the approach of constructing computational models of biological systems "executable biology", which focuses on the design of executable computer algorithms that mimic biological phenomena, an approach that the authors think must be integrated into biological research related to mathematical modeling.

However, defending paninformationalism and pancomputationalism can be good for two fundamental reasons. The first one relates to the centrality of natural human evolution, which presents that widespread activity of semiosis, I quoted above (including all kinds of signs, not only the propositional ones) that has been already generated by our ancestors at the times of the birth of the so-called "material culture". Indeed humans have constructed those huge *cognitive niches* (Odling-Smee et al. 2003; Laland and Sterelny 2006; Laland and Brown 2006), characterized by the presence of informational, cognitive (and more recently, computational) processes. This evolutionary evidence explains the philosophers' attitude to grant some ontological status to information and computation. However, I contend that adopting an approach in terms of "disseminated computation", such as the one I have proposed in the previous sections—the expression is introduced and explained above in Sect. 4.5.2—helps consider pancomputationalism in a naturalized way, acknowledging its limits and tempering its ontological and metaphysical overstatements.

In this chapter there is no room for dealing with the intricate jungle of the various concepts of pancomputationalism: the reader will find a detailed and rich description of various kinds of pancomputationalism, unlimited, limited, causal, interpretivist, ontic, etc. in Piccinini (2017).[25] Here it is sufficient to note that Horsman et al.'s hostility against pancomputationalism seems to refer to the "ontic" version of it, that is the idea that the universe itself is a computing system, and everything in it is a computing system too (or at least to the view that every physical system is a digital computing one and can be illustrated in computational terms).

We have to add that the strong advocates of the idea that universe computes contemplate a rather uncomplicated idea, and maintain that this happens thanks to different kinds of computation on different levels of organization, a contention that does not imply the acceptation that "everything is computation" (cf., for example, the already quoted info-computational view (Dodig-Crnkovic 2017)), which would instead become a vacuous statement equivalent to the claim that nothing is computation. In this perspective, they say, by introducing the new notion of universal computation where "universe" is a whole universe and not only the symbol manipulation logical device, an epistemological gain is reached: for example, we can build

---

[25] Pancomputationalism is also usefully discussed in the recent (Coelho Mollo, 2019) in the framework of an interesting and intricate comparison between the so-called "computational perspectivalism" and the mainstream accounts of physical computation, especially the teleologically-based mechanistic view.

new theoretical frameworks across disciplines that can help common understanding of the information processing in our brains and computations in the DNA or quantum computer. Grand unifications have always been attractive in physics and mathematics and they are not an obstacle to appreciating diversity in which real world unfolds for us. Again, the strong advocates of the idea that universe computes conclude that there would be absolutely no danger in broadening computational frameworks as much as it is justified by the behaviors of the systems that we study within the computational area of research.

After all, as contended by Kari and Rozenberg (2008) in the article "The many facets of natural computing", the most important property of computation consists in the multiplicity of its aspects and their efficacy. If a concept is capable of providing grand unification such as in the case of the concepts of matter, energy, information etc., it is worth to be used, developed, and explored. Finally, pancomputationalism is seen as a claim not critically dependent on the idea of discrete computation because the universe as an analog computer does not necessarily presuppose discrete computation. A main objection to this perspective is related to the following problem: one thing is modeling phenomena, systems, and data from a computational point of view, or exploiting new and unconventional "ignorant" substrates as computational tools, another is the adoption of the ontological idea that the objectivity itself has a computational nature (this last conviction leading to the so-called "ontic pancomputationalism" (Piccinini 2017). Analogously, as I have already said, one thing is to produce computational models for example of physical systems, another is to produce scientific knowledge of them.[26]

The second reason is cognitive and epistemological: in the literature of various disciplines often there is even an all-embracing concept of information, a kind of paninformationalism, in which physical (or biological) information is generalized to every state of a physical (or biological) system that is consequently defined as an information-pregnant state (Wolfram 2002; Lloyd 2006), which has to be considered relevant. Indeed, this approach has fecundated various excellent studies that physicists (for example Chiribella et al. 2012 and Goyal 2011), and some logicians, have realized, for instance developing mathematical frameworks for considering quantum theory in the light of the principles of information processing: a kind of information physics.

Finally, what about pancognitivism? An old-fashioned metaphysical theory, called synechism, contends that matter and mind are interrelated and in some sense it is not possible to differentiate them. The human cognitive mind would have germinated according to the same metaphysical laws that control the entire universe leading us to guess that the mind has an inclination (as Peirce contended) to find true abductive hypotheses regarding nature and the entire universe itself. As I have illustrated in Magnani (2009) it is Peirce himself that was convinced that "thought" is not only related to the brain: he said that it appears in the work of bees, crystals, and

[26] I have treated in detail this problem in Magnani (2018a).

throughout the purely physical world. In sum, seeing information, and/or cognition, and/or computation everywhere presents some virtues but some limitations have to be established, as I will sketch in the last subsection.

4.7.2 Protecting Ignorant Entities

Since the birth of the so-called *material culture* (Mithen 1996, 1999), when handaxes were made by Early Humans, the transformation of what I called ignorant entities in cognitive ones (and more recently in computational ones), also thanks to ancient techniques and modern technological advances, obviously had a tremendous positive impact on western civilization. However, as I have anticipated, a kind of warning has to be presented: I contend we are dealing in our era with what I call *overcomputationalization*. This means we are in the presence of too many entities and artifacts that carry computational tasks and powers. This implies some potential dangers: for example overcomputationalisation (1) furnishes too many opportunities to promote plenty of possible unresolvable disorganizational consequences in the activities of our societies and, (2) favors philosophical reflections that, taking advantage of the seduction exercised by old-fashioned metaphysical allures disguised as technical "analytical" studies, depict an oversimplified vision of the world.

There is not enough space here to deal with the first of the two issues above but I can suggest that a lot of research in human sciences has been produced to deal with the ethical consequences of our technological computational era.[27] Instead, I would like to say few words concerning the second issue: we face a decrease of entities that remain simply "ignorant" from the informational/cognitive but especially from the computational point of view. Buildings, streets, furniture, clothes, etc. became carriers not only of simple traditional informational/cognitive obvious qualities but also, thanks to computationalization, of further complicated propositional and model-based information, sometimes related to redundant and oppressive political, ideological, and economical aims.

Moreover, more and more engineered and artefactual ignorant entities, laptops, cell phones, servers, videos, cameras, social networks, robots, and also various unconventional substrates, intentionally computationally domesticated, as I have illustrated in this chapter, are invading human environments. The recent relevance acquired by the studies on the so-called Internet of Things is a clear indication of the importance of these processes (Smart et al. 2019). The excesses of overcompulationalism can also generate a decrease of attention to entities when simply seen as more or less "ignorant" (that is devoid of complicated cognitive and computational endowments) and for this reason they tend to appear more and more irrelevant. The same happens in the case of the excess of juxtaposition between new and complicated imposed informational/cognitive computational issues and various traditional entities. Unfortunately these effects (1) tend to cause, in the case of overcomputational-

[27] See for example my own book (Magnani 2007).

ism, too many constraints, limitations, and a weakening of human fruitful standard and creative (abductive) cognitive activities, as I have illustrated in the last chapter of my book (Magnani, 2017) and, (2) in the case of the excess of redundant cognitive/informational features computationally attributed to entities (features in this case exogenous to the original functions of those entities) they incline to impede to benefit from the simplification that would derive thanks to a reduction of the consequent cognitive and informational overload, often potentially and negatively distracting.

The reader does not have to misunderstand me, I believe that we should choose carefully what to defend from computationalization, and what is good to computationalize, and so my urgency to "protect ignorant entities" does not have to be intended as a warning about all kinds of computationalization. Computational modeling of human body and brain can for example fosters a better understanding of diseases, degenerative processes, aging and other processes of life, including processes in our brains that generate the human mind. Human brain is the largely unknown organ in the human body, and a lot of health problems, especially of aging populations is related to brains. Just to make few examples, we do not have cures for Alzheimer, Parkinson's disease or Autism. So fruitful current and future computational modeling will not only favor further domestication of *ignorant* entities by transforming them in "skilled" substrates able to carry computational and so cognitive performances, but also will fecundate scientific research related to the removal of another kind of *ignorance*, the classical one regarding the lack of understanding of high-level cognitive capacities both in human, non-human animals, and simpler organisms, and other biological entities such as cells, as also hoped by the admonition contained in the recent review "Time to reinspect the foundations? Questioning if computer science is outgrowing its traditional foundations" (Copeland et al. 2016).

4.7.3 The Future of Eco-Cognitive Settings: Computationally-Tailored or Humanly-Tailored?

In the previous chapters we have seen that data and types of problems that are basically privileged by current mainstream AI programs are very restricted, data preferred are quantity-related (Big Data) and the best strategies that can be simulated are the *locked* ones typical of well-established games. May be in the future AI will devote more attention to more circumscribed repositories of data (to become closer to the typical concrete conditions of human reasoning) and at the same time will be able to generate by itself new interesting information, to be in turn autonomously exploited, so increasing the openness of the range of data available and living together "in peace" with the human need of eco-cognitive openness.

A final ethical warning has to be clearly stressed: the widespread diffusion of standard and AI-based computational simulations will obviously incline human collectives to more and more adopt habits and settings that deal with data and problems appropriate to those same tools. This could generate an effect of progressive computational configuration of new social behaviors and environments appropriate to conform themselves to the computational tools and needs, to the detriment of old ones: in few words, will we face eco-cognitive settings which are *computationally-tailored*, characterized by the prevalence of locked cognitive strategies or *humanly-tailored*, characterized by the unlocked ones? As we have illustrated in the previous subsections the eco-cognitive openness in which human creative cognition dwells, could be reduced and jeopardized by the computationalization of the world, so threatening the chance of getting new and important knowledge results, and not only the scientific ones. Something similar could happen in the case of the prevalence of "locked"—in this case from the ethical point of view—eco-cognitive settings in which the same chance of making human ethical decisions will be threatened, because objectively impeded by computational machines embedded in a triumphant "computationally-tailored" eco-cognitive situation.

Summary

In this chapter, following the perspective I called "eco-cognitive computationalism", I have illustrated the new conceptual horizons generated by morphological computation as the effect of an unconventional and unexpected computationalization of bodies that are normally "ignorant". Morphological computation aims at the computational simplification of cognitive and motor tasks thanks to the building of smart "mimetic bodies" able to render an intertwined computation simpler, resorting to that "simplexity" of animal embodied cognition which represents one of the main quality of organic agents. Morphological computation can be seen as an activity of what I call "disseminated computation": the future of morphological computation research in robot design can lead to a new generation of robots endowed with improved adaptability and a minor quantity of involved control parameters. After having addressed the attention to the so-called "natural computing", to synthetic biology and DNA computing, and to the new perspectives of computational domestication of ignorant entities completely unconventional and/or unexpected, giving origin to new cognitive embodiments, the final part of the chapter presents a short discussion concerning the new concept of overcomputationalism. It is shown that the framework of disseminated computation helps us see the related concepts of pancognitivism, paninformationalism, and pancomputationalism in a more naturalized and prudent perspective, avoiding ontological or metaphysical overstatements. Finally, I have introduced a last question that clearly synthesizes the various problems illustrated in the chapter and that is presented to the reader as worth to be further discussed and studied: will the future of eco-cognitive settings computationally-tailored or humanly-tailored?

References

Ben Jacob, E., Shapira, Y., Tauber, A.I.: Seeking the foundation of cognition in bacteria. From Schrödinger's negative entropy to latent information. Phys. A **359**, 495–524 (2006)

Berthoz, A., Petit, J.L. (eds.): Complexité-simplexité. Collège de France, Paris (2014)

Burgin, M., Dodig-Crnkovic, G.: A taxonomy of computation and information architecture. In: Proceedings of the 2015 European Conference on Software Architecture Workshops, pp. 7:1–7:8. ACM, New York, NY, USA (2015)

Chiribella, G., D'Ariano, G.M., Perinotti, P.: Quantum theory, namely the pure and reversible theory of information. Entropy **14**, 1877–1893 (2012)

Clark, A.: Natural-Born Cyborgs: Minds, Technologies, and the Future of Human Intelligence. Oxford University Press, Oxford (2003)

Clark, A., Chalmers, D.J.: The extended mind. Analysis **58**, 10–23 (1998)

Coelho Mollo, D.: Against computational perspectivalism. Br. J. Philos. Sci. (2019) http://oup.prod.sis.lan/bjps/advancearticle-pdf/, https://doi.org/10.1093/bjps/axz036/29091662/axz036.pdf

Copeland, J., Dresner, E., Proudfoot, D., Shagrir, O.: Time to reinspect the foundations? Commun. ACM **59**(11), 34–38 (2016)

Crutchfield, J. P., Ditto, W. L., Sinha, S.: Introduction to focus issue: Intrinsic and designed computation: Information processing in dynamical systems—Beyond the digital hegemony (CHAOS, 20). Article number 037101 (2010)

Denning, P.J.: Computing is a natural science. Commun. ACM **50**(7), 14–18 (2007)

Denning, P.J., Martell, C.H.: Great Principles of Computing. The MIT Press, Cambridge, MA (2015)

Dodig-Crnkovic, G.: Significance of models of computation, from Turing model to natural computation. Mind. Mach. **21**(2), 301–322 (2011)

Dodig-Crnkovic, G.: The info-computational nature of morphological computing. In: Müller, V.C. (ed.) Philosophy and Theory of Artificial Intelligence, pp. 59–68. Springer, Berlin, Heidelberg (2013)

Dodig-Crnkovic, G.: Computational dynamics of natural information morphology, discretely/continuous. Philosophies **2**(4) (2017)

Fauconnier, G., Turner, M.: The Way We Think. Basic Books, New York (2003)

Fisher, J., Henzinger, T.A.: Executable cell biology. Nat. Biotechnol. **25**, 1239–1249 (2007)

Gil, R.: Minimal cell. In: Gargaud, M., Amils, R., Quintanilla, J.C., Cleaves, H.J., II., Irvine, W.M., Pinti, D.L., Viso, M. (eds.) Encyclopedia of Astrobiology, pp. 1065–1066. Springer, Berlin/Heidelberg (2011)

Goyal, P.: Information physics—Towards a new conception of physical reality. Information **3**, 567–594 (2011)

Grozinger, L., Amosand, M., Gorochowski, T.E., Carbonell, P., Oyarzún, D.A., Stoof, R., Goñi-Moreno, A.: Pathways to cellular supremacy in biocomputing. Nat. Commun. **10**. Article number 5250 (2019) http://www.nature.com/articles/s41467-019-13232-z

Hagiya, M., Aubert-Kato, N., Wang, S., Kobayashi, S.: Molecular computers for molecular robots as hybrid systems. Theor. Comput. Sci. **632**, 4–20 (2016)

Hauser, H., Füchslin, R.M., Nakajima, K.: Morphological computation—The physical body as a computational resource. In: Hauser, H., Füchslin, R.M., Pfeifer, R. (eds.) Morphological Computation: The Body as a Computational Resource, pp. 226–244. Self-published, Bristol (2014)

Hauser, H., Ijspeert, A.J., Füchslin, R.M., Pfeifer, R., Maass, W.: Towards a theoretical foundation for morphological computation with compliant bodies. Biol. Cybern. **105**(5), 355–370 (2011)

Horsman, C., Kendon, V., Stepney, S.: The natural science of computing. Commun. ACM **60**(8), 31–34 (2017a)

Horsman, C., Kendon, V., Stepney, S., Young, J.P.W.: Abstraction and representation in living organisms: When does a biological system compute? In: Dodig-Crnkovic, G., Giovagnoli, R. (eds.) Representation and Reality in Humans, Other Living Organisms and Intelligent Machines, pp. 91–116. Springer International Publishing, Cham (2017b)

Horsman, C., Stepney, S., Wagner, R.C., Kendon, V.: When does a physical system compute? Proc. R. Soc. A **470**, 1–25 (2014) (470 20140182; Published 9 July 2014). https://doi.org/10.1098/rspa.2014.0182

Hutchins, E.: Material anchors for conceptual blends. J. Pragmat. **37**, 1555–1577 (2005)

Hutchison III, C., Chuang, R.Y., Noskov, V., Assad-Garcia, N., Deerinck, T., Ellisman, M., Venter, J.: Design and synthesis of a minimal bacterial genome. Science **351**, aad6253-1–aad6253-11 (2016)

Kari, L.: DNA computing based on insertions and deletions (2013) (oai:CiteSeerX.psu:10.1.1.297.3308)

Kari, L., Rozenberg, G.: The many facets of natural computing. Commun. ACM **51**(10), 72–83 (2008)

Kari, L., Seki, S., & Sosík, P.: DNA computing: Foundations and implications. In: Rozenberg, G., Bäck, T., Kok, J. (eds.) Handbook of Natural Computing, Vol. 3, pp. 1073–1128. Springer, Cham, Switzerland (2012)

Kluger, J.: Simplexity. Why Simple Things Become Complex (and How Complex Things Can Be Made Simple). Hyperion Books, New York, NY (2008)

Knuuttila, T., Loettgers, A.: Mathematization in synthetic biology: Analogies, templates, and fictions. In: Lenhard, J., Carrier, M. (eds.) Mathematics as a Tool: Tracing New Roles of Mathematics in the Sciences, pp. 37–56. Springer, Cham (2017)

Laland, K.N., Brown, G.R.: Niche construction, human behavior, and the adaptive-lag hypothesis. Evol. Anthropol. **15**, 95–104 (2006)

Laland, K.N., Sterelny, K.: Perspective: Seven reasons (not) to neglect niche construction Evolution. Int. J. Org. Evol. **60**(9), 4757–4779 (2006)

Lloyd, S.: Programming the Universe: A Quantum Computer Scientist Takes on the Cosmos. Knopf, New York, NY (2006)

Longo, G.: Critique of computational reason in the natural sciences. In: Gelenbe, E., Kahane, J.-P. (eds.) Fundamental Concepts in Computer Science. Imperial College Press/World Scientific, London (2009a)

Longo, G.: Turing and the "imitation game" impossible geometry: Randomness, determinism and programs in Turing's test. In: Epstein, R., Roberts, G., Beber, G. (eds.) Parsing the Turing Test: Philosophical and Methodological Issues in the Quest for the Thinking Computer, pp. 377–411. Springer, Dordrecht (2009b)

Longo, G.: Incomputability in physics and biology. Math. Struct. Comput. Sci. **22**(5), 880–900 (2012)

Longo, G.: The biological consequences of the computational world: Mathematical reflections on cancer biology (2017). arXiv:1701.08085

Magnani, L.: Abduction, Reason, and Science: Processes of Discovery and Explanation. Kluwer Academic/Plenum Publishers, New York (2001)

Magnani, L.: Morality in a Technological World. Knowledge as Duty. Cambridge University Press, Cambridge (2007)

Magnani, L.: Abductive Cognition. The Epistemological and Eco-Cognitive Dimensions of Hypothetical Reasoning. Springer, Heidelberg/Berlin (2009)

Magnani, L.: Understanding Violence. The Intertwining of Morality, Religion, and Violence. A Philosophical Stance. Springer, Heidelberg/Berlin (2011)

Magnani, L.: The Abductive Structure of Scientific Creativity. An Essay on the Ecology of Cognition. Springer, Cham, Switzerland (2017)

Magnani, L.: Computationalism in a dynamic and distributed eco-cognitive perspective. In: Dodig-Crnkovic, G., Burgin, M. (eds.) Philosophy and Methodology of Information, vol. 1, pp. 365–288. World Scientific, Singapore (2018a)

Magnani, L.: Eco-cognitive computationalism: From mimetic minds to morphology-based enhancement of mimetic bodies. Entropy **20**(6) (2018b)

Maturana, H.R.: Reality: The search for objectivity. Irish J. Psychol. **9**, 25–82 (1988)

Maturana, H.R., Varela, F.J.: Autopoiesis and Cognition: The Realization of the Living. Reidel, Dordrecht (1980)

Maturana, H.R., Varela, F.J.: The Tree of Knowledge: The Biological Roots of Human Understanding. Shambhala, Boston, MA (1987)

Mithen, S.: The Prehistory of the Mind: A Search for the Origins of Art, Religion, and Science. Thames and Hudson, London (1996)

Mithen, S.: Handaxes and ice age carvings: Hard evidence for the evolution of consciousness. In: Hameroff, A. R., Kaszniak, A.W., Chalmers, D.J. (eds.), Toward a Science of Consciousness III: The Third Tucson Discussions and Debates, pp. 281–296. MIT Press, Cambridge (1999)

Müller, V.C., Hoffmann, M.: What is morphological computation? On how the body contributes to cognition and control. Artif. Life **23**(1), 1–24 (2017)

Nakajima, K., Hauser, H., Kang, R., Guglielmino, E., Caldwell, D., Pfeifer, R.: A soft body as a reservoir: Case studies in a dynamic model of octopus-inspired soft robotic arm. Front. Comput. Neurosci. **7**, 91 (2013)

Nakajima, K., Hauser, H., Li, T., Pfeifer, R.: Information processing via physical soft body. Sci. Rep. **5** (2015)

Odling-Smee, F.J., Laland, K.N., Feldman, M.W.: Niche Construction: The Neglected Process in Evolution. Princeton University Press, Princeton, NJ (2003)

Peirce, C.S.: Collected Papers of Charles Sanders Peirce, vols. 1–6, Hartshorne, C., Weiss, P. (eds.); vols. 7–8, Burks, A.W. (ed.) (1931–1958). Harvard University Press, Cambridge, MA (1866–1913)

Piccinini, G.: Computation in physical systems. In: Zalta, E.N. (ed.) The Stanford Encyclopedia of Philosophy (Summer 2017). Stanford University, Metaphysics Research Lab (2017)

Rozenberg, G., Bäck, T., Kok, J. (eds.): Handbook of Natural Computing. Springer, Cham, Switzerland (2012)

Smart, P., Madaan, A., Hall, W.: Where the smart things are: Social machines and the internet of things. Phenomenol. Cogn. Sci. **18**(3), 551–575 (2019)

Steward, J.: Cognition=life: Implications for higher-level cognition. Behav. Proc. **35**, 311–326 (1996)

Thom, R.: Esquisse d'une sémiophysique. In: Meyer, V. (Trans). Semio Physics: A Sketch, Addison Wesley, Redwood City, CA, 1990. InterEditions, Paris (1988)

Turing, A.M.: The chemical basis of morphogenesis. Philos. Trans. R. Soc. **237**(641), 37–72 (1952)

Turing, A.M.: Intelligent machinery [1948]. In: Meltzer, B., Michie, D. (eds.) Machine Intelligence, Vol. 5, pp. 3–23. Edinburgh University Press, Edinburgh (1969)

Vallverdú i Segura, J.: Computational epistemology and e-science: A new way of thinking. Minds Mach. **19**(4), 557–567 (2009)

Varela, F.J.: Principles of Biological Autonomy. Elsevier, New York (1979)

Varela, F.J.: Patterns of life: Intertwining identity and cognition. Brain Cogn. **34**, 72–87 (1997)

Wolfram, S.: A New Kind of Science. Wolfram Media, Champaign (2002)

Yaman, F., Adler, A.: How can AI help Synthetic Biology? In: AAAI Synthetic Biology (2013)

Yaman, F., Adler, A., Beal, J.: AI challenges in Synthetic Biology engineering. In The Thirtieth AAAI Conference on Innovative Applications of Artificial Intelligence (IAAI-18) (2018)

Conclusion

In place of a formal conclusion, I offer here a sort of final comment about what I see as the most important elements of *Eco-Cognitive Computationalism*. I hope that I was able, through my arguments, to encourage a serious commitment to increasing our knowledge about the importance of seeing the eco-cognitive aspects of the current trends of computation, as a philosophical, epistemological, cognitive, and ethical topic. The analysis that I tried to operate, focused on the examination of computation *as it is* and *as it should become*, as I abstained from explaining it *away* by any reduction to a rigid unilateral perspective, can improve our chances to maintain the theoretical commitment towards the study of computation as a multifarious phenomenon, deeply affecting human beings in their individual and social lives, but also from the perspective of their cognitive processes and chances.

At the same time, given the importance of computation in actual human behavior, studies on this topic help to maintain an intellectual focus on moral commitments, hopefully, also beyond the strictly intellectual community. The knowledge about computation I intensely endorsed in this book aims also at supplying researchers with the philosophical poise required to handle controversial issues regarding the intertwining between computation, current AI research, unconventional tendencies, what I called the overwhelming domestication of "ignorant entities", also taking advantage of a new reexamination of some seminal ideas contained in Turing's article "Intelligent Machinery", of 1948.

Standard and traditional computation, increasing cognitive domestication of ignorant entities through computation, and what I called overcomputationalization, are so pervasive, so much a part of human life, that it is not easy to imagine how to live without being affected by their effects. The strong impact computational world bear on human beings, but also animals and nature in general, through its entanglement with other technologies, depends on complicated cognitive, biological, and epistemological mechanisms, which can be seen and further analyzed only by taking advantage of an eco-cognitive framework, as I tried to demonstrate in this book. People can make use of different computational resources as weapons of better cognitive

L. Magnani, *Eco-Cognitive Computationalism*, Cognitive Systems Monographs 43,
https://doi.org/10.1007/978-3-030-81447-2

activities, but also to harm others—both for defensive and aggressive scopes—as I for example described along with the various chapters of this book when emphasizing the role of computers as mimetic minds but also when dealing with the uncertain future of eco-cognitive settings, easily imaginable as computationally rather than humanly tailored. Stressing these issues has been a further extension of the analysis of computation, transcending the precious and sophisticated traditional examination of its foundations, acknowledging our "condition" of "cyborgs", to adopt Andy Clark's expression (2003), as a way of accepting our responsibility when we are handling in various ways the various kinds of computation. It could hopefully be of help in individuating cognitive firewalls that prevent overcompulationalization from always escalating, to further weaken damages and to improve our human abductive creativity, freedom, and safer ownership of our destinies.

In sum, understanding computation in an eco-cognitive perspective can be of help, for example, to avoid falling into a situation when humans become passive receivers of effects generated by computational machines, with consequent damage to the ownership of their destiny. Seeing computation from the perspective of the neutral moral commitment of its developers—where computation appears, clearly, so positively value-driven—can help to grasp computation both as a common and at this point ineluctable event, which cannot be disregarded by academicians and that it is worth being examined taking advantage of a clear intertwining between the epistemological and the ethical perspective, avoiding separate treatments incapable of fruitful communicating each other. To understand eco-cognitive computationalism I had to contain my own personal moral and epistemological predilections, but, for example, I have gained a new philosophical awareness about the current human condition in front of overcomputationalization. The spectacle of computationalization taught me that we have to live with it and we have to monitor it taking advantage of a wide interdisciplinary approach that needs further deep and audacious new knowledge.

Let us begin.

The manufacturer's authorised representative in the EU is Springer Nature Customer Service Centre GmbH, Europaplatz 3, 69115 Heidelberg, Germany. If you have any concerns regarding our products, please contact ProductSafety@springernature.com

Printed and bound by CPI Group (UK) Ltd, Croydon, CR0 4YY
30/09/2026
02225608-0007